Renewable Resources for Biorefineries

RSC Green Chemistry

Editor-in-Chief:
Professor James Clark, *Department of Chemistry, University of York, UK*

Series Editors:
Professor George A Kraus, *Department of Chemistry, Iowa State University, Ames, Iowa, USA*
Professor dr. ir. Andrzej Stankiewicz, *Delft University of Technology, The Netherlands*
Professor Peter Siedl, *Federal University of Rio de Janeiro, Brazil*
Professor Yuan Kou, *Peking University, China*

How to obtain future titles on publication:
A standing order plan is available for this series. A standing order will bring
delivery of each new volume immediately on publication.

For further information please contact:
Book Sales Department, Royal Society of Chemistry, Thomas Graham House,
Science Park, Milton Road, Cambridge, CB4 0WF, UK
Telephone: +44 (0)1223 420066, Fax: +44 (0)1223 420247
Email: booksales@rsc.org
Visit our website at www.rsc.org/books

Renewable Resources for Biorefineries

Edited by

Carol Sze Ki Lin
City University of Hong Kong, Hong Kong
Email: carollin@cityu.edu.hk

Rafael Luque
University of Cordoba, Spain
Email: q62alsor@uco.es

THE QUEEN'S AWARDS
FOR ENTERPRISE:
INTERNATIONAL TRADE
2013

RSC Green Chemistry No. 27

Print ISBN: 978-1-84973-898-9
PDF eISBN: 978-1-78262-018-1
ISSN: 1757-7039

A catalogue record for this book is available from the British Library

Published by The Royal Society of Chemistry,
Thomas Graham House, Science Park, Milton Road,
Cambridge CB4 0WF, UK

Registered Charity Number 207890

For further information see our web site at www.rsc.org

Preface

With an increasing awareness and concerns about our dependency on fossil resources and the depletion of crude oil reserves, this book aims to give an overview to the bio-based economy. Experts from academia and industries worldwide have presented their views on industrial biotechnology, green chemistry and sustainable policy related to the use of renewable raw materials for non-food applications and energy supply. Innovative key research concepts such as waste valorization and its society aspects related to renewable resources need to be transferred to the public, academic and industrial sectors in order to redouble efforts towards a more sustainable society and a bio-based economy.

This book is a concise presentation of a variety of such important aspects in renewable resources and biorefineries. Key features of the book include guidelines for appropriate practices on biorefineries and understanding of waste valorization and bio-processing, with updated latest information on international research and development of novel green strategies and technologies for utilization of renewable resources. The book contains eight high quality chapters, detailing both theoretical and practical information with the aim to provide inspiration for additional research and application to generate energy and value-added products. Bio-based polymers and materials, downstream processing of bioplastic materials, enhanced biomass degradation by polysaccharide monooxygenases, microalgae technology, application of food waste valorization technology in Hong Kong, advanced generation of bioenergy, pre-treatment and transformation of ligno-cellulosics, and bioactive compounds from biomass are topics extensively covered in this monograph.

As editors of this book, we sincerely hope it can serve as a starting point to highlight the importance of industrial biotechnology and green chemical processes and technologies in biorefineries for the benefit of humankind.

RSC Green Chemistry No. 27
Renewable Resources for Biorefineries
Edited by Carol Sze Ki Lin and Rafael Luque
© The Royal Society of Chemistry 2014
Published by the Royal Society of Chemistry, www.rsc.org

We would also like to take this opportunity to acknowledge and thank all the contributors to this book for their excellent collaboration and timely contributions which brought together a highly comprehensive range of topics. Many thanks to Merlin Fox, Commissioning Editor for the Green Chemistry Series at Royal Society of Chemistry (RSC) and the production team at RSC, particularly Helen Prasad, for her helpful assistance through this project and ensuring everything came together properly.

With very best wishes for successful and enjoyable reading.

Carol Sze Ki Lin
Rafael Luque

Contents

RSC Green Chemistry No. 27
Renewable Resources for Biorefineries
Edited by Carol Sze Ki Lin and Rafael Luque
© The Royal Society of Chemistry 2014
Published by the Royal Society of Chemistry, www.rsc.org

CHAPTER 1

Bio-based Polymers and Materials

NATHALIE BEREZINA[*a] AND SILVIA MARIA MARTELLI[b]

[a] Materia Nova, Rue des Foudriers 1, 7822 Ghislenghien, Belgium;
[b] Faculty of Engineering, Federal University of Grande Dourados,
Dourados, Brazil
*Email: nathalie.berezina@materianova.be

1.1 Introduction

Biomaterials have gained attractiveness in the last decades due to both
ecological and economic concerns. Increased pollution, and especially
visible pollution, has first driven the scientific and industrial communities
to look at biofragmentable and biodegradable substitutes for traditional
petroleum-based non-biodegradable materials. Then the dramatic increase
of oil prices before the economic crisis of 2007 influenced the move from the
biodegradable to the bio-based. Finally, the compliance of the obtained
materials with thermo-mechanical constraints has turned interest to the
partially bio-based materials.

Bio-based materials can be obtained mainly by two different ways: the
direct production of polymers or the production of bio-based monomers and
their further (bio)chemical polymerization. The direct production of biopolymers can be achieved by microorganisms (polyhydroxyalkanoates, PHA),
by algae (alginate *etc.*), by superior plants (pectin *etc.*) or by several types of
producers, *e.g.* cellulose is produced by superior plants but also by bacteria,
chitosan is produced by crustacean but also by fungi.

RSC Green Chemistry No. 27
Renewable Resources for Biorefineries
Edited by Carol Sze Ki Lin and Rafael Luque
© The Royal Society of Chemistry 2014
Published by the Royal Society of Chemistry, www.rsc.org

Whatever the producer of biopolymers, the main difficulty is to trigger its composition. Indeed, the obtained material has to comply with the thermo-mechanical constraints of its anticipated usages, and these characteristics are strongly related to the monomeric composition of the polymer and its size. This common problem of biopolymers does not have a unique solution. In the case of PHA, several microorganisms can produce the same polymer and the modification of the feeding substrates influences the monomeric composition of the polymers produced by the same microorganism, whereas important differences are found in the monomeric composition of the poly/oligo-saccharides produced by different (micro)organisms. The regulation of the size of the biopolymers appears even more complicated; although PHA can be obtained with important masses, polysaccharides are actually mainly oligosaccharides. In any case, the main difficulty consists in making the (micro)organism perform the biosynthesis we want consistently and repeatedly.

To circumvent this problem, one can be tempted to make more controlled chemical polymerization with the bio-based monomers. Thus the production of bio-based monomers was also developed. However, the polymerization of bio-based monomers often asks for more development, as in the case of polylactic acid (PLA) and of polybutylene succinate (PBS); moreover, the thermo-mechanical needs for the expected applications are hardly reached with these polymers. Therefore, two more options can be foreseen: the production of partially bio-based materials (Sorona®) or the production of bio-monomers identical to the already existing and improved petroleum-based (ethylene, isobutylene, caprolactam *etc.*).

In this chapter we discuss the main biomaterials produced by these different methods as well as the achieved improvements and remaining bottlenecks in their production, modifications and applications.

1.2   Direct Production of Biopolymers

1.2.1   PHA

Polyhydroxyalkanoates (PHA) were discovered in 1926 by Maurice Lemoigne[1] as energy storage materials in *Bacillus megatherium* and *Bacillus mesentericus vulgatis*. Still, they had to wait until the 1960s and for *Cupriavidus* genera (previously referred as *Hydrogenomonas, Alcaligenes, Ralstonia* and *Wautersia*) to be extensively studied.[2-4] Indeed, the accumulation of the PHA by this genera has appeared to be more effective. Moreover, the first petrol crisis and further ecological issues increased the awareness and the interest in the bio-based materials.

Several PHA producing microorganisms as well as several types of PHA were discovered.[5-8] The whole PHA family can be sub-divided into three main categories: the short-chain length PHA (PHA$_{SCL}$), the medium-chain length PHA (PHA$_{MCL}$) and rarer PHA (Figure 1.1). The structural differences inside the PHA family imply deep differences in their thermo-mechanical

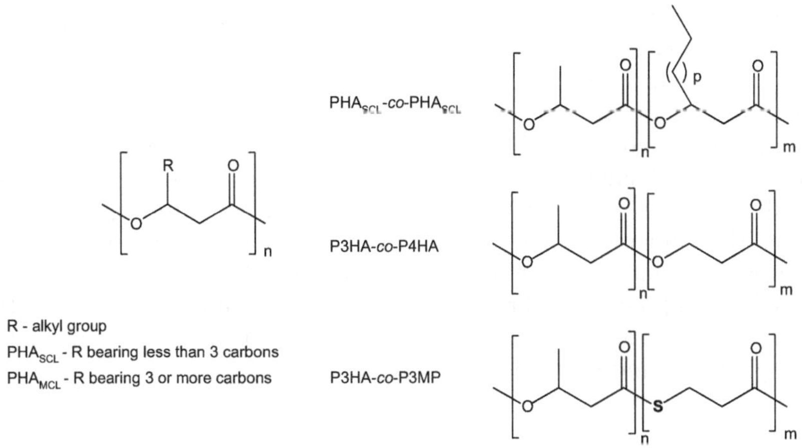

Figure 1.1   Major types of PHA found in the nature.

properties. Thus, PHA_{SCL} mainly composed by polyhydroxybutanoates (PHB) and poly(hydroxybutanoate-*co*-valerate) (PHBV), are crystalline polymers, which are rather brittle and stiff, with high melting points (near 160–180 °C) and low glass transition temperature (between −5 and 0 °C), whereas the PHA_{MCL} are thermoplastic elastomers with low crystallinity and tensile strength with high elongation to break (400–700%).[9,10]

1.2.1.1 PHA_{SCL}

PHA_{SCL} are the most studied biopolymers among the PHA family. Numerous improvements of their production have been achieved during these last decades. These improvements mainly concerned the selection of wild-type strains (121 g L^{-1} of PHB was thus achieved using *Cupriavidus* genera[11]), the engineering of strains (161 g L^{-1} of PHB was reported with *E. coli* (XL1-Blue) strain[11]), the feeding strategy 'nutrient limited' versus 'nutrient sufficient' conditions,[12,13] with the latter having been recently proved to be most efficient for the main producing strains (33 times productivity enhancement[14]).

Also, the growth and the PHA accumulation on wastes and by-products have been paid important attention in recent years, in order to enhance the economic and sustainable efficiencies. Thus, different alternative substrates were tested – such as vinasse,[15] oil palm frond juice,[16] soybean oil,[17] waste glycerol[18] and other by-products from the biodiesel industry.[19,20] Unfortunately up to now these strategies have not shown comparable productivities as an artificial carbon source (only 67.2 g L^{-1} of PHB were produced when soybean oil was used as a substrate).

Even if the *Cupriavidus* genera remains predominant in the production of PHA_{SCL}, other genera were also discovered and studied in recent years: *Bacillus cereus*,[21] *Brevundimonas vesicularis*,[22] *Sphingopyxis macrogoltabia*,[18]

Nostoc muscorum,[23] *Synechocystis* sp.,[24] *Herbaspirillum seropedicae*,[25] *Haloferax mediterranei*[26] etc.

The important issues of the control of the biopolymer composition, the relative abundance of the 3-hydroxybutanoate (3-HB) and 3-hydroxyvalerate (3-HV), was also addressed by different feeding strategies, namely the choice of the 3-HV inducing substrates[27-30] (up to 80% of 3-HV content was obtained with 1 g L^{-1} mixture of levulinic acid and sodium propionate[31]) as well as the choice of one-time initial versus sequential addition of those substrates, the latter been found more efficient.[32-35]

Finally, bioprocess improvements such as solid-state fermentation (SSF),[36,37] continuous and two-stage culture systems,[38] down-stream processing (DSP)[39] and purification[40] were studied.

1.2.1.2   PHA$_{MCL}$

PHA$_{MCL}$ are mainly produced by the *Pseudomonads*.[41] They are usually synthesized as copolymers of two or three or even more monomers, obtained by β-oxidation of fatty acids used as feeding substrates, the monomeric parts usually bear $n \pm 2$ carbons. One noticeable exception to this general rule is the recently reported *Pseudomonas mendocina* strain able to produce pure homopolymers of poly-3-hydroxyoctanoates (PHO).[42]

Although several strategies, such as multiple nutrient limitation,[43,44] batch and chemostat strategies[45,46] or strain engineering,[47-49] were attempted for improving the PHA$_{MCL}$ productivity, it remains rather low compared to the results of PHA$_{SCL}$. Thus PHA production of 0.2 g L^{-1} h^{-1} was observed for *Pseudomonas oleovorans* grown on octanoic acid,[50] or 47% of PHA inside the cells of *Pseudomonas putida* grown on 11-phenoxydecanoic acid, or 53–58% of the conversion of raw materials by *Comomonas testosterone* grown on vegetable oil.[51]

1.2.1.3   Rarer PHA

With PHA$_{MCL}$ presenting more interesting thermo-mechanical properties and PHA$_{SCL}$ being more easily produced it was tempting to try to combine the advantages by synthesizing the PHA$_{SCL}$-*co*-PHA$_{MCL}$. The most studied among these copolymers is poly(3-hydroxybutanoate-*co*-3-hydroxyhexanoate) (P(HB-*co*-HHx)). *Aeromonas caviae* seems to be one of the rare bacteria to naturally produce such copolymers, the main results being obtained with engineered strains. The best results so far (up to 70% of 3-HHx content) was obtained with the *Cupriviadus necator* engineered with the *Rhodococcus aetherivorans* PHA synthase, grown on crude kernel oil.

Other rare PHA are mainly composed of P3HA-*co*-P4HA[52-54] and the thiopolyesters polyhydroxybutanoate-*co*-polymercaptoprionate (PHB-*co*-PMP),[55-57] however, the whole PHA family contains more than hundred different polymers, and is still growing.[58,59]

1.2.1.4 Applications and Industrial Production of PHA

Applications of PHA have evolved. Initially foreseen applications in packaging have been recently replaced by more promising and cost-compatible medical applications. Numerous devices (patches, scaffolds *etc.*), wound management tools (suture, dressings), drugs delivery systems and pro-drugs were based on these biopolymers.[60–63]

The observed shift in the applications of PHA has significant importance on the production of those polymers. More particularly the high purity required for the final products designed for medical application may not be compatible with the use of wastes and by-products as raw materials. Thus, in the near future we will observe a shift from the study of cheap raw materials (in order to lower the overall cost of PHA[64–66]) to purification processes in order to separate the PHA from the enzymes and proteins linked to the PHA granules inside the cells.[67 69]

Although present worldwide, the total industrial production of PHA remains tiny (Table 1.1). However, very recently one of the main historical producers of PHA, Metabolix, has achieved for the first time a $3.6 million profit in 2011.[70] It now becomes reasonable to foresee more success stories in the future for this family of biopolymers.

1.2.2 Polysaccharides and Oligosaccharides

Polysaccharides and oligosaccharides are widely produced in nature. Animals and plants are the most important producers in terms of volume, whereas microorganisms produce much wider diversity. Also, the small-scale structure, the proportion of different sugars and the type of linkage play an important role in the final properties of these biopolymers. Thus, applications of these poly- and oligo-saccharides are also very different, from low- to high-added value products.

Table 1.1 Main PHA producers worldwide.

Company	Polymer type	Announced capacity	Geographical localization	Ref.
Newlight Technologies	PHA	100 000 lbs per year	Southern California (USA)	177
Meridian (DaniMer)	PHA	300 000 tons per year (announced capacity) (15 000 tons per year actual production)	Bainbridge, Georgia (USA)	177, 178
Metabolix (Antibioticos)	PHA (Mirel)	10 000 tons per year	Leon (Spain)	177
Biomer	PHB	1 000 tons per year	Krailig (Germany)	179
Bio-on	PHA (Minerv)	10 000 tons per year	Bologne (Italy)	180, 181
Ecomann	PHA	not specified	Shenzhen (China)	
GreenBio	PHA	10 000 tons per year	Tianjin (China)	

1.2.2.1 Polysaccharides Produced by Plants

Plants are the most important producers of polysaccharides. Their main products are cellulose, hemicellulose, starch, inulin and pectin. Cellulose is by far the most abundant renewable polymer available worldwide, its occurrence was estimated at some 10^{11}–10^{12} tons per annum.[71]

Cellulose and starch are both homopolymers, only composed by D-glucose units (Figure 1.2). The only difference between them is the type of linkage between the sugar units. Cellulose is composed by β-D-glucose units, whereas starch has α-D-glucose units. This tiny structural difference makes a huge difference in the properties of these two polysaccharides. Starch is digestible product used by many organisms on earth, whereas the digestion of cellulose is very difficult as it is often requiring both physical and chemical steps. Starch, used by humans for centuries for food and feed (as well for its nutritional value as thickener and emulsifier),[72] has found application also in textile and paper industries and as a biodegradable packaging material.[73–75] More recently it has been investigated as a first-generation bio-fuel, but raised ethical issues due to the famine problem in different parts of the world. On the other hand, the traditional usage of cellulose was in the paper and textile industries,[74] but more technical applications have been recently explored using nanocrystals[76] or grafted cellulosic materials.[77–79] Nowadays it is also abundantly studied as a second-generation bio-fuel.

Inulin is another homopolymer produced by plants. It is composed mainly of fructose units, even if a starting glucose moiety can be present (Figure 1.3). The main producers of inulin are either chicory and artichokes or biocatalytically synthesized fructo-oligosaccharides.[80] Inulin is mainly used in the food industry for both its nutritional and technological advantages. Indeed, being composed of fructose units, inulin is hardly hydrolysed during the digestion process.[81,82] The non-food applications of inulin have also been recently investigated; until now they concern merely modified inulin, thus carboxymethylinulin (CMI) was successfully used as dispersing agent and dicarboxyinulin (DCI) as a builder or co-builder in detergent formulation to replace polyacrylates.[83]

Pectins are polysaccharides bearing methyl-esterified galacturonic acid and rhamnose units.[83] The proportions of these units as well as the presence of other constitutive units depend on the plant from which the studied pectin was isolated.[84] Pectins are biosynthetically produced in the Golgi apparatus of plants. Although many plants are able to produce pectins, industrial pectin is mainly extracted from citrus peel and apple pomace under mildly acidic conditions. The main applications of pectins remain in the food industry as gelling agents, while more technical applications could arise from more homogeneous polysaccharides obtained by chemical modification or gene technology.[85,88]

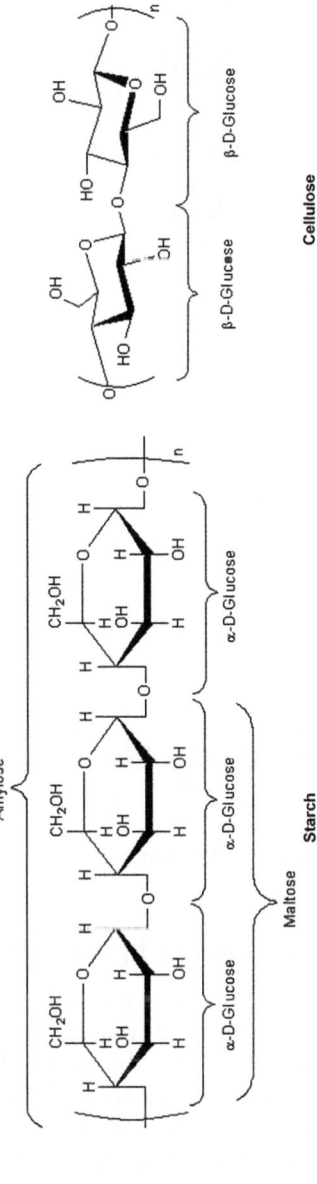

Figure 1.2 Structure of starch and cellulose.

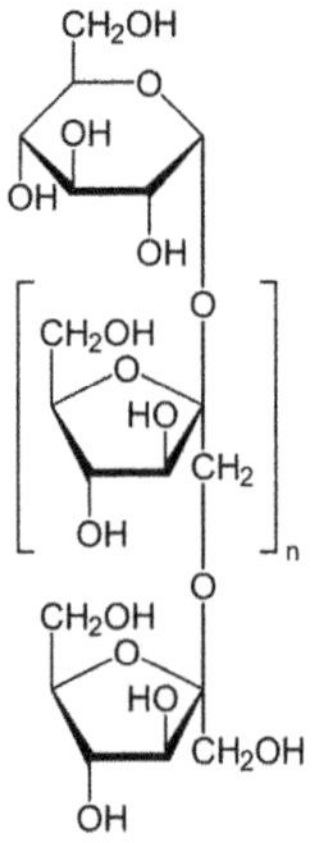

Figure 1.3   Structure of inulin.

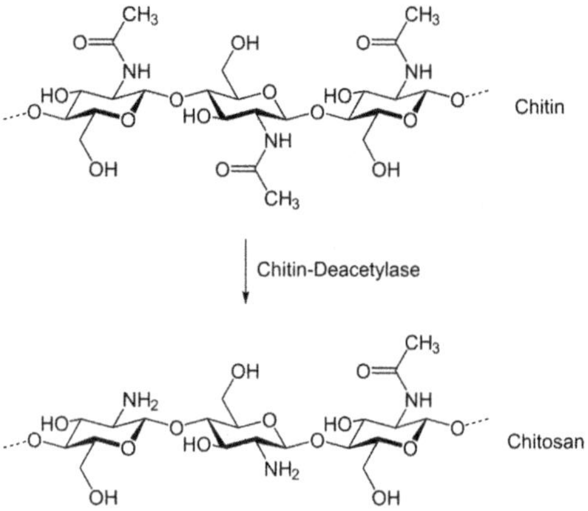

Figure 1.4   Structure of chitin and chitosan.

1.2.2.2   *Polysaccharides Produced by Animals*

Chitin and chitosan are the main polysaccharides produced by animals. Those molecules are part of insects' and crustaceans' exoskeletons,[86] even if some mushrooms and fungi[87] are also able to produce them. Chitin and chitosan are aminoglucopyranans composed of *N*-acetylglucosamine (GlcNAc) and glucosamine (GlcN) units (Figure 1.4). Currently the most important source of chitin and chitosan remains chemical processing of the waste fraction of the shellfish industry, even if some biotechnological and entomological studies are in progress. The main applications of these polysaccharides are, until now, based on their antimicrobial properties as applied to either the food or the cosmetic industries.[88]

1.2.2.3 Polysaccharides Produced by Microorganisms

Compared to higher plants and animals, microorganisms are characterized by a wide diversity of the poly- and oligo-saccharides they produce (Table 1.2). Among the microorganisms, bacteria have the widest spread of possibilities. Polysaccharides produced by bacteria are mainly extracellular polysaccharides (EPS),[89] also called exopolysaccharides, whereas those produced by algae are mainly cell wall and structural constituents.

Intracellular bacterial polysaccharides have not yet found proper applications; they are, however, extensively studied as storage materials similar to those in humans (glycogen[90]) and as specific targets for the drug attacks of pathogens (murein,[91] teichoic and teichonic acids[92]).

The main applications of EPS are in the food and cosmetic industries as thickeners, gelling agents and emulsifiers or in pharmacy and medicine,[93–95] with some of them being used as the active principles (schizophyllan[96]).

Some of the microbial polysaccharides are also produced by higher organisms, such as cellulose in plants or chitin and chitosan in animals; however, microbial production is often better controlled and offers the possibility of higher-added value applications.[89,97]

Among other applications we can underline are the use as oil-drilling agents (dextran and derivatives,[98] xanthan,[99] sphingan,[100] scleroglucan[101]), immobilization supports (curdlan,[102] alginate[103]), cement-based materials (sphingan[99]), adhesives (pullulan[104]) and can sealing (alginate[105]).

1.2.3 Others

Several other biopolymers are directly synthesized in nature, such as proteins, poly(amino acid)s, lignin, humic substances or sporopollenin. Until now, they are under-used and under-studied compared to the previously detailed two main families of directly produced biopolymers. Thus in this paragraph we will only mention some recent developments in the field of proteins and poly(amino acid)s, and lignin.

1.2.3.1 Proteins and Poly(amino acid)s

Proteins are composed of amino acids linked by peptide bonds. Two main biosynthetic pathways for protein production have been identified so far: the ribosomal and the non-ribosomal–multi-enzyme paths.[106] The proteins are mainly heteropolymers composed by a variety of amino acids; however, three poly(amino acid)s can be obtained through the multi-enzyme pathway: cyanophycin (aspartic acid–arginine dipeptide), ε-poly-L-lysine and poly-α,β-aspartic acid (Figure 1.5).

The main applications of proteins are in the nutraceutical[107] and pharmaceutical[108] industries. The specific antimicrobial properties of ε-poly-L-lysine promoted its utilization in the food industry in Japan,[109] whereas poly-α,β-aspartic acid is mainly used as a polydispersant in detergents.[110]

Table 1.2 Main polysaccharides produced by the microorganisms.

Type of microorganism	Polysaccharide	Type	Main producing strain	Application	Ref.
Bacteria	cellulose	extracellular	*Acetobacter, Rhizobium, Rhizobacterium, Agrobacterium, Sarcina*	paper, textile, food, cosmetics, medicine	96
	curdlan	extracellular	*Alcaligenes, Agrobacterium*	food, pharmaceutical, agricultural, support for immobilization	101
	dextran and derivatives	extracellular	*Streptococcus, Leuconostoc*	oil drilling, food, agriculture	97
	hyaluronan and hyaluronic acid	extracellular	*Streptococcus, Pasteurella*	cosmetics, medicine	182
	xanthan	extracellular	*Xanthomonas*	food, oil drilling	98
	glycogen	storage (accumulated)	*E. coli, Clostridia, Bacillus, Streptomyces*		92
	succinoglycan	extracellular	*Rhizobium, Agrobacterium, Alcaligenes, Pseudomonas*	thickening, gel-forming, precipitation agent	183
	alginate	extracellular	*Pseudomonas, Azotobacter*	food, pharmaceutical, biotechnology (immobilization)	102
	glucuronan	extracellular	*Rhizobium, Pseudomonas*	cosmetics, agriculture, medicine	184
	sphingan	extracellular	*Sphingomonas*	food, biotechnology (solid culture media & gel electrophoresis), construction (cement-based materials), oil drilling fluids	99

Type	Polymer	Location	Organism	Application	Ref
	alternan	extracellular	*Leuconostoc*	cosmetics, food, pharmaceuticals	185
	levan	extracellular	*Bacillus, Zymomonas, Aerobacter, Pseudomonas*		186
	murein	cytoplasm	any		93
	teichoic and teichuronic acids	cell wall	Gram-positive bacteria		94
Fungi	pullulan	extracellular	*Aureobasidium, Pullularia, Dematium*	food, pharmaceuticals, industry (adhesives)	103
	chitin/chitosan	cell wall	*Basidomycetes, Ascomycetes, Phycomycetes/Mucorales*	absorption of coloring matters & metal, medicine	89
	scleroglucan	extracellular	*Sclerotium*	food, medicine, oil drilling	100
	schizophyllan (sizofilan, sizofiran)	extracellular	*Schizophyllum*	medicine (anti-tumor)	95
Algae	alginate	structural	*Phaeophyceae* (brown algae)	shear-thinning viscosifyer for textile, paper coating, can sealing, medicine, pharmacy, food	104
	carragenan	cell wall	*Rhodophyceae* (red seaweeds)	gelling, thickening, stabilizing agents	187
	ulvan	cell wall	*Ulva* sp.	food, pharmaceuticals	188

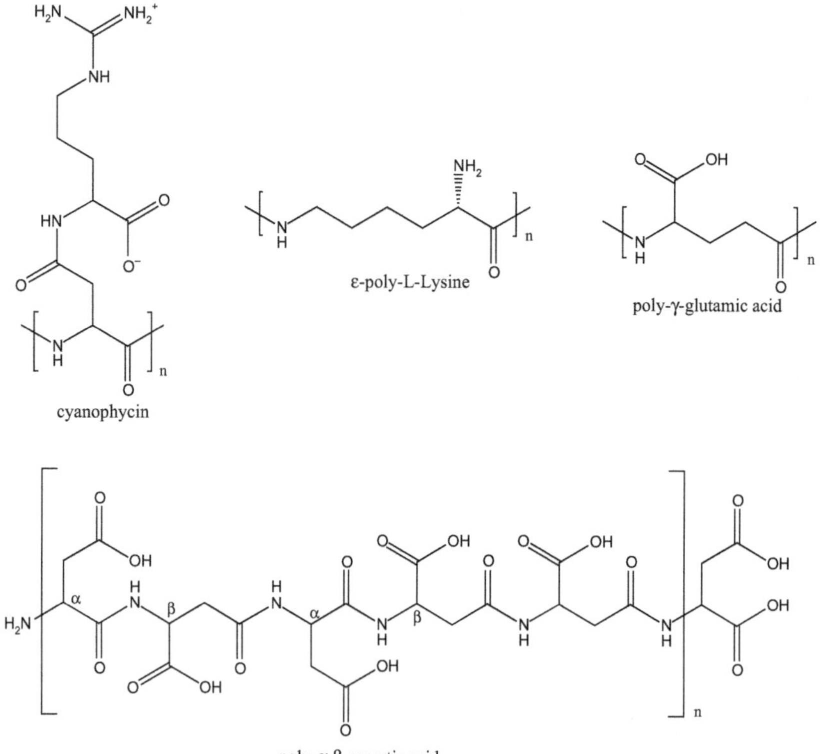

Figure 1.5   Molecular structure of poly(amino acid)s.

Some proteins were studied for a long time for their materials applications: soy protein,[111] wheat gluten[120] and collagen (the denatured form being called gelatine).[112,113] In all the cases, the stability of the proteins and their sensitivity to moisture require strengthening by plasticization, compatibilization, cross-linkage[114,115] or production of protein–nanoclay composites.[116]

1.2.3.2   Lignin

Lignin is, besides cellulose and hemicellulose, the third main constituent of plants. Lignin possesses a very complex, cross-linked, polyphenolic structure.[117] Despite its really important chemical potential as the sole abundant source of bio-based aromatic compounds, the inherent difficulties of the purification and homogenization of lignin severely limits its widespread usage.[118]

Current studies also cover the purification[119,120] and the de- and repolymerization of lignin,[121–123] as its application in the materials industry.[124] In this fast-evolving context, the drawback is that the price of lignin, previously merely considered as waste or by-product, rose, whereas valuable applications are not yet clearly identified.[125,126]

1.3 Production of Bio-based Monomers for Further Polymerization

In the category of the polymers produced from bio-based monomers, the polyesters used to be more popular. Thus, historically, the main studied monomers were bi-functional molecules, such as lactic acid, an α-hydroxy acid able to self-condense for the production of polylactic acid (PLA); 1.3-propanediol (PDO) leading to Dupont's Sorona® after condensation with terephthalic acid and succinic acid, foreseen to be a key bio-based building block and leading to polybutylene succinate (PBS) after condensation with 1,4-butanediol.

More recently, some other monomers have been studied. The attraction of production of well-known materials from renewable feedstock led to studies of the use of ethanol for the production of bio-based polyethylene (PE), of caprolactam and muconic acid for the production of polyamides (PA) and of isobutylene for the synthesis of polyisobutylene.

1.3.1 Lactic Acid and PLA

Lactic acid used to be an important molecule for the chemical and food industries for centuries. It is produced through anaerobic fermentation by many bacteria. Traditionally its main applications are in the food industry where it is used as a natural acidifying agent.[127,128] More recently, the scope of its applications was significantly enlarged by the synthesis of polylactic acid (PLA) as a new biodegradable and bio-based bioplastic.[129]

The synthesis of PLA from lactic acid cannot really be achieved by simple condensation, which mainly leads instead to the oligomers. Therefore a more original way of synthesizing PLA was achieved by ring-opening polymerization (ROP) starting from the lactide, the lactic acid dimer (Figure 1.6), with the lactide itself being obtained by the partial de-polymerization of the oligomers.[130]

Lactic acid bears an asymmetric carbon and is mainly produced by bacteria at nearly enantiopure (S) (L) configuration. The improvement of thermo-mechanical properties of PLA and more specifically of its heat resistance require the combination of pure P(L)LA and P(D)LA in so-called stereocomplexes.[131] The production of the (R)-lactic acid was therefore studied, even if until now the strains producing this enantiomer are rarely reported and the production remains lower than for the (S) counterpart.[132]

Recycling currently appears to be a more sustainable approach for the management of the end of life of PLA than composting. Thus, de- and re-polymerization techniques were studied.[133] These studies have faced

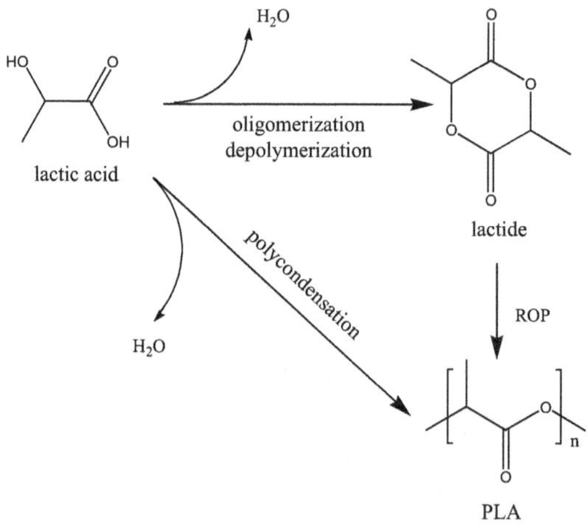

Figure 1.6   Synthesis of PLA.

the racemization issue,[134,135] therefore desymmetrization approaches were studied. The stereoselective oxidation to pyruvic acid unfortunately leads to the loss of 50% of the desired product[136,137] or to the reduction of the same enantiomer of the lactic acid.[138] Biocatalytic discrimination using the *Candida antarctica* lipase B (CALB) remained unsuccessful.[139] Finally, kinetic resolution using (*R*)-myrtenol or the separation of the diastereoisomers obtained with (*S*)-2-methylbutanol as a chiral auxiliary[140] were found to be the most promising at the current stage.

   PLA suffers from its cost and also some weaknesses in thermo-mechanical properties (mainly heat resistance); however, it is considered as one of the most promising bioplastics for the substitution of the petroleum-based polymers in materials and packaging applications.[141–143] The main industrial PLA companies are Nature Works, invested by Cargill and PTT Global Chemicals, with the announced production of 140 000 metric tons of PLA per year; as well as the traditional lactic acid producers such as Purac through Synbra and the collaboration with Sulzer Chemtec and Galactic through its joint venture with Total, Futerro (1500 tons per year).

1.3.2   1,3-Propanediol and Sorona®

1,3-Propanediol (PDO) is one of the oldest known products of anaerobic fermentation. Numerous wild-type genera, *Klebsiella*,[144] *Enterobacter*,[145] *Citrobacter*,[154] *Clostridium*,[146] *Lactobacillus*,[157] convert glycerol to PDO (Figure 1.7). Most of them are class II pathogens, *i.e.* opportunistic pathogens. Also the production of PDO by wild-type strains leads to the concomitant production of several by-products, such as 2,3-butanediol, lactate, acetate, formate, ethanol *etc.* Finally, despite the assumption of the imminent over-production of glycerol due to the extensive use of biodiesel, PDO remains more expensive than glucose.

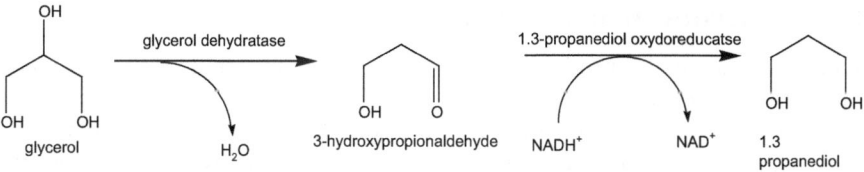

Figure 1.7 Metabolic pathway for the production of 1,3-propanediol from glycerol.

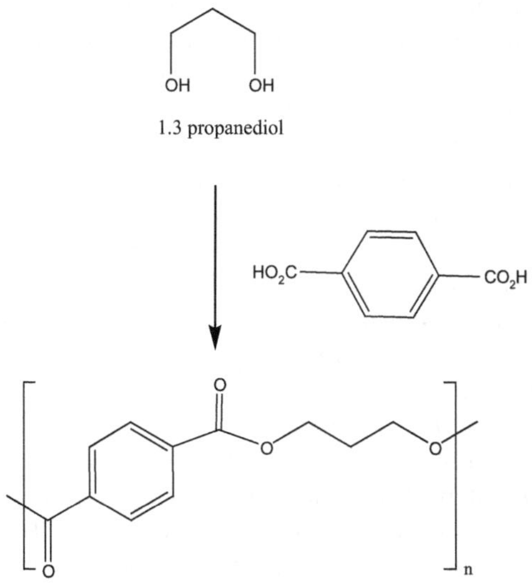

Figure 1.8 Synthesis of Sorona®.

Several studies were accomplished in order to achieve production of PDO by engineered strains. Two different strategies were attempted: expression of genes for the glycerol production from glucose in PDO-producing strains and the expression of the PDO-producing genes in the glycerol-producing strains. The first strategy was found to be more efficient and has been applied industrially by Genencor,[147–149] a Danisco's subsidiary.

Further, DuPont performed the polymerization of PDO with terephthalic acid (Figure 1.8) to obtain Sorona®.[150]

This partially bio-based polymer is mainly used in different fibre applications such as floor coverings and sportswear. The PDO success story is considered as an important milestone from both industrial and societal points of view because it led to the investigation of other possibilities for the coupling of bio-based and traditional monomers (see below for PBS) with further consideration of the importance of the bio-based content in polymeric materials (controlled by the ASTM D 6866 norm, for example), and to the ultimate acquisition of Danisco with its Genencor subsidiary by DuPont in 2011.[151]

1.3.3 Succinic Acid and PBS

Succinic acid can be produced by anaerobic fermentation of several wild-type strains: *Anaerobiospirillum*,[152] *Propionibacterium*,[153] *Escherichia*,[154] *Pectinas*. Succinic acid is considered as one of the major bio-based platform chemicals (Figure 1.9). Its applications are foreseen as well in commodity chemicals (THF, hydroxysuccinimide *etc.*), as organic acids (malic, fumaric, itaconic *etc.*) or polymers (polybutylene succinate, PBS).[155]

Therefore, several academic and industrial studies were launched for the improvement of the production of succinic acid. Several enzymes involved in the metabolic pathway for the production of succinic acid, fumarate reductase, PEP carboxylase, malate deshydratase, were cloned and over-expressed in *E. coli* or in *S. cerevisiae*; improvements and mutations of wild strains were also performed.

Currently, the main industrial programs aiming at the production of succinic acid are headed at 2000 ton-scale by BioAmber (a joint venture of ARD and DNP), at 100 000 ton-scale by Reverdia (a joint venture of Roquette and DSM) and at 60 000 ton-scale by Succinity (a joint venture of Purac and BASF). However, the current market of succinic acid (produced from fossil raw materials) is only 15 000 tons per year. The industrial companies are clearly anticipating the enhancement of the demand for bio-based succinic acid, mainly due to the PBS applications.

The production of PBS remains mainly at the research and development stages. The direct condensation of succinic acid and 1,4-butanediol encounters the same problems as in the PLA case discussed above. To circumvent these issues, different strategies were adopted: the two-stage melt polycondensation (esterification and polycondensation), the condensation–extension approach using hexamethylene diisocyanate as extension agent. This latter strategy gave the most promising results until now with the M_n and MW of 40 000 and 100 000 g mol^{-1} respectively.[156–158] The use of cyclic monomers was yet poorly explored and mainly used in enzymatic polymerizations.[159–161]

1.3.4 Others

1.3.4.1 Ethanol and PE

Ethanol is one of the oldest biotechnological products used by humans, even ancient Egyptians were drinking a sort of beer obtained by alcoholic fermentation. More recently ethanol was involved in first- and then second-generation biofuels. The overall sustainability and economic viability of these approaches remain doubtful. However, the production of bio-ethanol in important amounts led to its consideration for bio-based plastics production. Thus, the dehydration of ethanol was extensively studied and improved to produce ethylene, while the further polymerization to polyethylene (PE) and utilization are well known in the plastic industry.[162] Furthermore, bio-ethanol was also used for the production of partially bio-based polyethylene terephthalate (PET).[163]

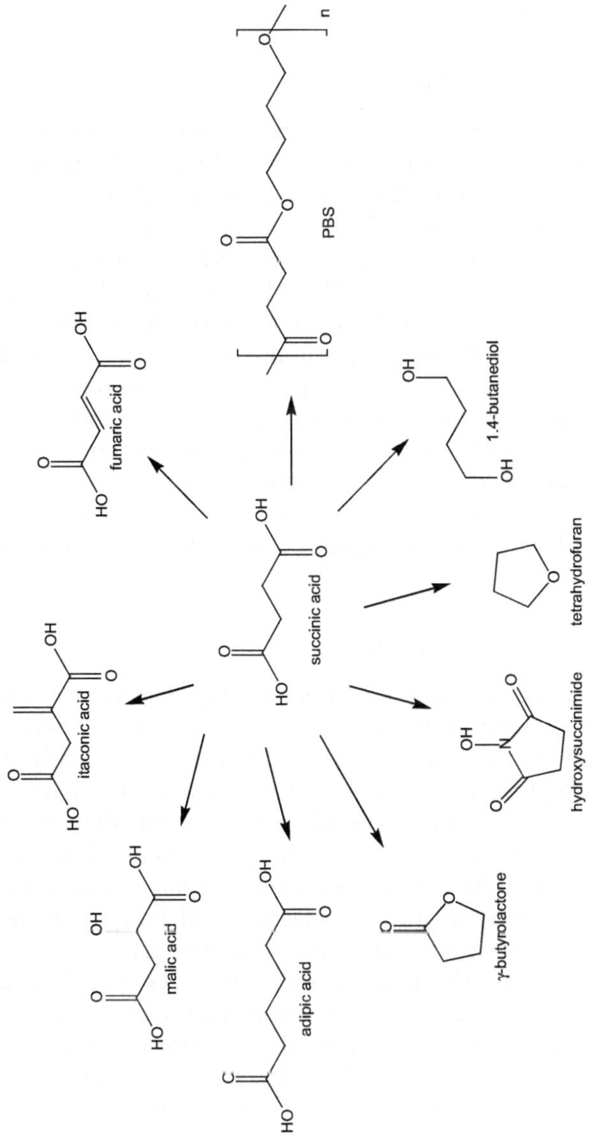

Figure 1.9 Scope of applications of succinic acid.

Braskem is currently the main actor for the production of bio-PE.[164] The Braskem's bio-PE is produced from sugar cane ethanol; in order to satisfy its production capacity of 200 000 tons per year, it uses arable land estimated at 0.02% of all the available land in Brazil. Further developments in bio-based polyolefin concern bio-polypropylene (PP) obtained through the metathesis of the ethylene dimer.

1.3.4.2   Caprolactam and PA-6

The polyamide-6 (PA-6) was first synthesized from caprolactam by Wallace Carothers in 1935 when working for DuPont. Then, it became one of the most popular polymers worldwide, reaching an annual production of nearly 2 billion tons. Its main producer is the Dutch DSM. Thus it is understandable that this company was pioneering the biotechnological approach for the production of bio-caprolactam from the α-ketoglutarate (Figure 1.10).

It is worth noticing the retrosynthetical approach of this synthesis, as the first patented step was the cyclization of 6-aminocaproic acid to caprolactam,[165] followed by the transformation of α-ketopimelate into 6-aminocaproic acid[166] and finally the elongation of α-ketoglutarate to the α-ketopimelate.[167] This latter step is of particular interest, as it uses the iteration of acetyl-CoA's addition–dehydration–hydrogenation–decarboxylation cascade ending by the addition of one carbon atom to the main skeleton, and can be followed until α-ketosuberate (three carbons) have been added to the initial chain (Figure 1.11).

1.3.4.3   Adipic Acid and PA-6,6

Adipic acid is the most important commercial aliphatic dicarboxylic acid. Its main application is the synthesis of the polyamide-6,6 (PA-6,6), another important polyamide first synthesized by Carothers in the early 1930s. The direct biotechnological access to adipic acid from sugars remains challenging. It is hypothesized that 2-oxoadipate, an intermediate in the L-lysine biosynthesis, could be used; however, the subsequent elimination of the keto-group still needs to be confirmed. Besides, it can also be obtained from the degradation of cyclohexane, caprolactam and long-chain dicarboxylic acids, or nitrile hydrolysis of adiponitrile[168] (Figure 1.12).

The uncertainty about direct biotechnological access to adipic acid offers a route to the exploration of possible biosynthesis of its intermediates – glucaric and muconic acids.[169,171] The final conversion of these intermediates to adipic acid is until now performed only by chemical means. An interesting point is the possibility to produce the muconic acid from aromatics,[170] thus combining depollution and the synthesis of useful chemicals.

1.3.4.4   Isobutylene and Polyisobutylene

Until now, the bio-based monomers were composed by hydrocarbons bearing oxygen or nitrogen heteroatoms, with pure hydrocarbons being only

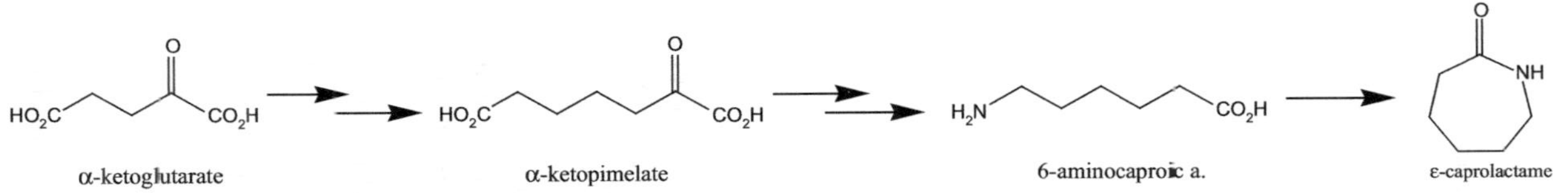

Figure 1.10 Biotechnological route for the production of caprolactam.

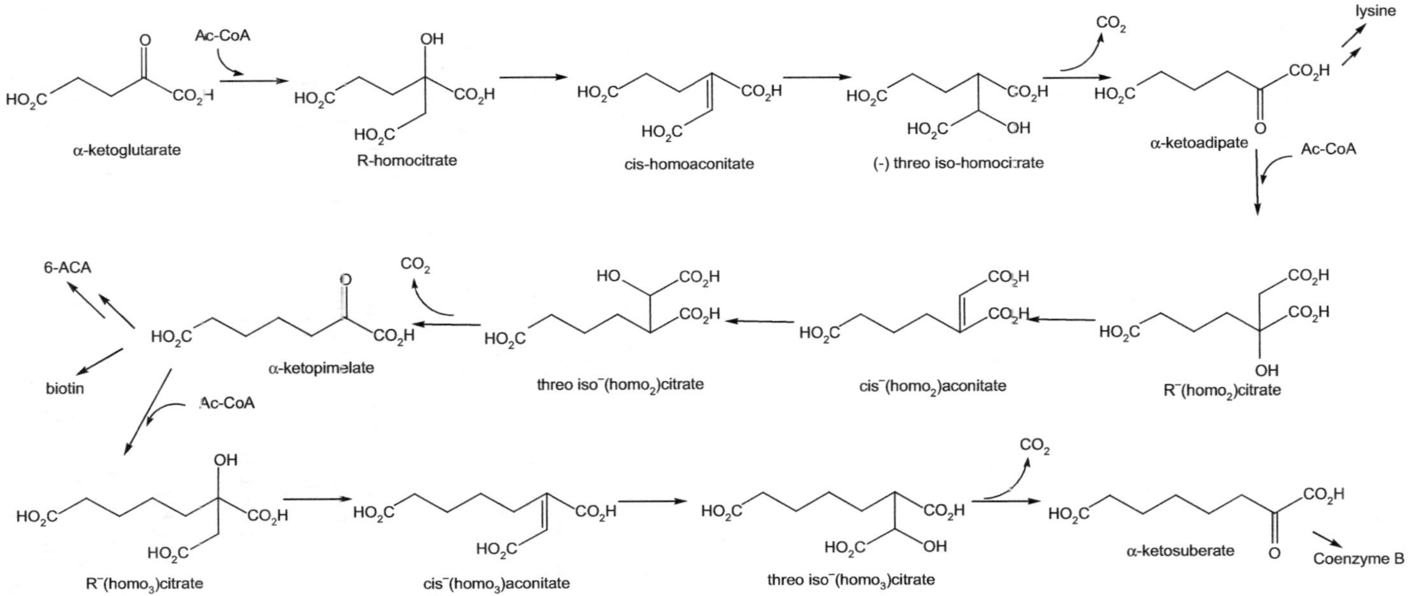

Figure 1.11 The engineered pathway for the production of α-ketosuberate from α-ketoglutarate.

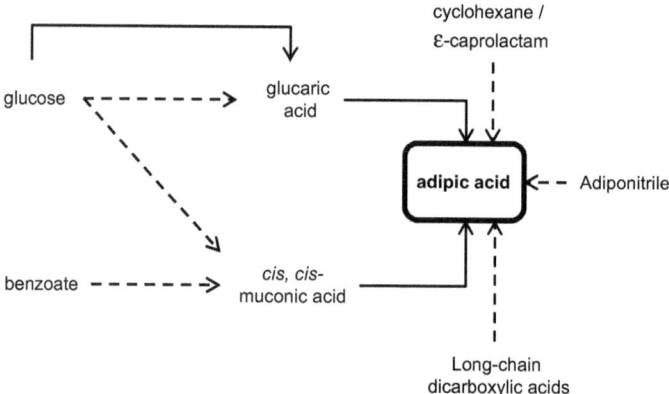

Figure 1.12   Biotechnological routes for the production of the adipic acid.

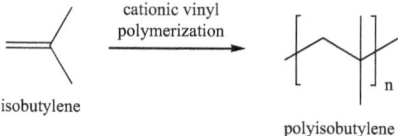

Figure 1.13   Polymerization of isobutylene.

obtained from fossil feedstock. This assertion is no longer true. French biotechnological company Global Bioenergies has recently patented a biotechnological process for the production of isobutylene, the monomer leading to the polyisobutylene widely used for its gas barrier properties (Figure 1.13). Besides the impressive technological breakthrough of this process, whose economic feasibility has yet to be discussed, the described procedure is very elegant, as the product is recovered in gas form from the reaction media, thus avoiding its saturation.

Further developments of pure hydrocarbon monomers leading to bio-isoprene are currently under development by Ajinomoto (teaming with Bridgestone), DuPont (teaming with Goodyear), Glycos biotechnologies and Aemetis. Bio-based butadiene is currently under development by Genomatica and Global Bioenergies in collaboration with Lanza Tech.[171]

1.4   Outlook

After the academic community's fixation on biodegradable products, bio-based products became more attractive. Recycling appears to be more sustainable than composting as an end-of-life solution. This twist is extremely important for future research and industry directions. Thus, low-added value applications, such as packaging, are hardly considered any longer. Biomaterials whose main attractiveness was based for years on bio-degradability and did not show any particular thermo-mechanical properties (PLA) are currently under extensive research for more high-added value applications.[172–175]

The difficulty of producing tailor-made 100% bio-based biomaterials (PHA, polysaccharides *etc.*) led to the investigation of partially bio-based versions of already well-established petroleum-based polymers (PET) or the development of new ones (PBS, Sorona®). It is worth noticing that even if at the current stage of knowledge these materials look like a tiny compromise between sustainability and economic efficiency, the development of genetic tools allows hope in further full bio-based versions of these materials.[176]

Recent breakthroughs also concern the development of pure bio-based products identical to the petroleum-based materials (PE, PP, PA-6, PA-6,6, polyisobutylene *etc.*). The economic outcomes of these initiatives are clearly dependent on the global energy supply. The recent enhancement of shale gas production, including the production of so-called 'wet gases' such as butane, propane and ethane, will deeply influence the economic per-spectives of bio-PE and bio-PP.[179]

Finally, neither price nor pure ideals can be considered as valuable driving forces in the present-day ever-evolving globalized world. The intrinsic per-formance and special characteristics of the materials could be more valuable guidelines for the mid-term research and investments.

References

1. M. Lemoigne, *Bull. Soc. Chim. Biol.*, 1926, **8**, 770–782 (in French).
2. E. Schuster and H. G. Schlegel, *Arch. Microbiol.*, 1967, 380–409 (in German).
3. H. G. Schlegel and R. Lafferty, *Nature*, 1965, **205**, 308–309.
4. A. A. Chowdhury, *Arch. Microbiol.*, 1963, **47**, 167–200 (in German).
5. P. Y. Tian, L. Shang, H. Ren, Y. Mi, D. D. Fan and M. Jiang, *African J. Biotechnol.*, 2009, **8**, 709–714.
6. P. Suriyamongkol, R. Weselake, S. Narine, M. Moloney and S. Shah, *Biotechnol. Adv.*, 2007, **25**, 148–175.
7. S. Y. Lee, *Tibtech*, 1996, **14**, 431–438.
8. R. Amache, A. Sukan, M. Safari, I. Roy and Keshavarz, *Chem. Eng. Trans.*, 2013, **32**, 931–936.
9. E. Akaraonye, T. Keshavarz and I. Roy, *J. Chem. Technol. Biotechnol.*, 2010, **85**, 732–743.
10. S. Khanna and A. K. Srivastava, *Process Biochem.*, 2005, **40**, 607–619.
11. Y. Tokiwa and C. U. Ugwu, *J. Biotechnol.*, 2007, **132**, 264–272.
12. E. J. Borman *et al.*, *Appl. Microbiol. Biotechnol.*, 1998, **50**, 604–607.
13. O. P. Peoples and A. J. Sinskey, *J. Biol. Chem.*, 1989, **264**, 15298–15303.
14. N. Berezina, *New Biotechnol.*, 2013, **30**, 192–195.
15. A. Bhattacharyya, A. Pramanik, S. K. Maji, S. Haldar, U. K. Mukhopadhyay and J. Mekherjee, *AMB Express*, 2012, **2**, 1–10.

16. M. A. K. M. Zahari, H. Ariffin, M. N. Mokhtar, J. Salihon, Y. Shirai and M. A. Hassan, *J. Biomed. Biotechnol.*, 2012, DOI: 10.1155/2012/125865.
17. J. G. Cruz Pradella, J. L. Ienczak, C. R. Delgado and M. K. Taciro, *Biotechnol. Lett.*, 2012, **34**, 1003–1007.
18. J. M. B. T. Cavalheiro, R. S. Raposo, M. C. M. D. Almeida, M. T. Cesario, C. Sevrin, C. Grandfils and M. M. R. Fonseca, *Bioresour. Technol.*, 2012, **111**, 391–397.
19. I. L. Garcia, J. A. Lopez, M. P. Dorado, N. Kopsahelis, M. Alexandri, S. Papalikolaou, M. A. Villar and A. A. Koutinas, *Bioresour. Technol.*, 2013, **130**, 16–22.
20. I. V. Spoljaric, M. Lopar, M. Koller, A. Muhr, A. Salerno, A. Reiterer, K. Malli, H. Angerer, K. Strohmeier, S. Schober, M. Mittelbach and P. Horvat, *Bioresour. Technol.*, 2013, **133**, 482–494.
21. S. P. Valappil, S. K. Misra, A. R. Boccaccini, T. Keshavarz, C. Bucke and I. Roy, *J. Biotechnol.*, 2007, **132**, 251–258.
22. J. A. Silva, L. M. Tobella, J. Becerra, F. Godoy and M. A. Martinez, *J. Biosci. Bioeng.*, 2007, **103**, 542–546.
23. L. Sharma, A. K. Singh, B. Panda and N. Mallick, *Bioresour. Technol.*, 2007, **98**, 987–993.
24. B. Panda, P. Jain, L. Sharma and N. Mallick, *Bioresour. Technol.*, 2006, **97**, 1296–1301.
25. A. I. Catalan, F. Ferreira, P. R. Gill and S. Batista, *Enz. Microbiol. Technol.*, 2007, **40**, 1352–1357.
26. D. Zhao, L. Cai, J. Wu, M. Li, H. Liu, J. Han, J. Zhou and H. Xiang, *Appl. Microbiol. Biotechnol.*, 2013, **97**, 3027–3036.
27. K. Sudesh, H. Abe and Y. Doi, *Prog. Polym. Sci.*, 2000, **25**, 1503–1555.
28. M. Zinn, H. U. Weilmann, R. Hany and M. Schmid, *et al.*, *Acta Biotechnol.*, 2003, **23**, 309–316.
29. W. H. Lee, C. Y. Loo, C. T. Nomura and K. Sudesh, *Bioresour. Technol.*, 2008, **99**, 6844–6851.
30. I. S. Sankhla, R. Bhati, A. K. Singh and N. Mallick, *Bioresour. Technol.*, 2010, **101**, 1947–1953.
31. N. Berezina, *Biotechnol. J*, 2012, 7, 304–309.
32. A. S. Kelley, N. V. Mantzaris, P. Daoutidis and F. Srienc, *Nano Lett.*, 2004, **1**, 481–485.
33. E. N. Pederson, C. W. J. mc Chalicher and F. Srienc, *Biomacromolecules*, 2006, 7, 1904–1911.
34. N. V. Mantzaris, A. S. Kelley and F. Srienc, *AIChE J*, 2001, **47**, 727–743.
35. N. V. Mantzaris, A. S. Kelley, P. Daoutidis and F. Srienc, *Chem. Eng. Sci.*, 2002, **57**, 4643–4663.
36. F. C. Oliveira, *et al.*, *Bioresour. Technol.*, 2007, **98**, 633–638.
37. L. R. Castilho, *et al.*, *Bioresour. Technol.*, 2009, **100**, 5996–6009.
38. G. Du *et al.*, *J. Biotechnol.*, 2001, **88**, 59–65.
39. S. P. Valappil *et al.*, *J. Biotechnol.*, 2007, **132**, 251–258.
40. A. Althuri *et al.*, *Bioresour. Technol.*, 2013, **145**, 290–296.

41. B. Kessler and N. J. Palleroni, *Int. J. Syst. Evol. Microbiol.*, 2000, **50** 711–713.

42. R. Rai, D. M. Yunos, A. R. Boccaccini, J. C. Knowles, I. A. Barker, S. M. Howdle, G. D. Tredwell, T. Keshavarz and I. Roy, *Biomacromolecules*, 2011, **12**, 2126–2136.

43. T. Egli and M. Zinn, *Biotechnol. Adv.*, 2003, **22**, 35–43.

44. M. Zinn, B. Witholt and T. Egli, *J. Biotechnol.*, 2004, **113**, 263–279.

45. R. Durner, M. Zinn, B. Witholt and T. Egli, *Biotechnol. Bioeng.*, 2001, **72**, 278–288.

46. R. Hartmann, R. Hany, E. Pletscher, A. Ritter, B. Witholt and M. Zinn, *Biotechnol. Bioeng.*, 2006, **93**, 737–746.

47. Y. Poirier, N. Erard and J. MacDonald-Comber Petetot, *Appl. Environ. Microbiol.*, 2001, **67**, 5254–5260.

48. B. Zhang, R. Carlson and F. Srienc, *Appl. Environ. Microbiol.*, 2006, **72**, 536–543.

49. L. Cai, M. Q. Yuan, F. Liu and G. Q. Chen, *Bioresour. Technol.*, 2009, **100**, 2265–2270.

50. L. J. R. Foster, R. A. Russel, V. Sanguanchaipaiwong, D. J. M. Stone, J. M. Hook and P. J. Holden, *Biomacromolecules*, 2006, 7, 1344–1349.

51. N. Thakor, U. Trivedi and K. C. Patel, *Bioresour. Technol.*, 2005, **96**, 1843–1850.

52. H. Abe and Y. Doi, in *Biopolymers – Polyesters II*, Wiley-VCH, 2002, pp. 105–133.

53. J. M. B. T. Cavalheiro, M. C. M. D. Almeida, M. M. R. Fonseca and C. C. C. R. Carvalho, *J. Biotechnol.*, 2012, **164**, 309–317.

54. D. H. Park and B. S. Kim, *New Biotechnol*, 2011, **28**, 719–724.

55. S. Tanaka, L. Feng and Y. Inoue, *Polym. J.*, 2004, **36**, 570–573.

56. T. Lutke-Eversloh, K. Bergander, H. Luftmann and A. Steinbuchel, *Microbiology*, 2001, **147**, 11–19.

57. T. Lutke-Eversloh and A. Steinbuchel, *FEMS Microbiol. Lett.*, 2003, **221**, 191–196.

58. Y. Takagi, R. Yasuda, A. Maehara and T. Yamane, *Eur. Polym. J.*, 2004, **40**, 1551–1557.

59. P. Phukon, B. Pokhrel, B. K. Konwar and S. K. Dolui, *Biotechnol. Lett.*, 2013, **35**, 607–611.

60. T. Keshavarz and I. Roy, *Curr. Opin. Microbiol.*, 2010, **13**, 321–326.

61. S. Philip, T. Keshavarz and I. Roy, *J. Chem. Technol. Biotechnol.*, 2007, **82**, 233–247.

62. M. Zinn, B. Witholt and T. Egli, *Adv. Drug Deliv. Rev.*, 2001, **53**, 5–21.

63. C. Vilos, F. A. Morales, P. A. Solar, N. S. Herrera, F. D. Gonzales-Nilo, D. A. Aguayo, H. L. Mendoza, J. Comer, M. L. Bravo, P. A. Gonzalez, S. Kato, M. A. Cuello, C. Alonso, E. J. Bravo, E. I. Bustamante, G. I. Owen and L. A. Velasquez, *Biomaterials*, 2013, **34**, 4098–4108.

64. C. Du, J. Sabirova, W. Soetaert and S. K. C. Lin, *Curr. Chem. Biol.*, 2012, **6**, 14–25.

65. T. M. Keenan, J. P. Nakas and S. W. Tanenbaum, *J. Ind. Microbiol. Biotechnol.*, 2006, **33**, 616–626.
66. T. Tsuge, *J. Biosci. Bioeng.*, 2002, **94**, 579–584.
67. N. Thomson, I. Roy, D. Summers and E. Sivaniah, *J. Chem. Technol. Biotechnol.*, 2010, **85**, 760–767.
68. N. Jacquel, C. W. Lo, Y. H. Wei and H. S. Wu *et al.*, *Biochem. Eng. J.*, 2008, **39**, 15–27.
69. A. Maehara, S. Taguchi, T. Nishiyama, T. Yamane and Y. Doi, *J. Bacteriol.*, 2002, **184**, 3992–4002.
70. *Bioplastic World,* April 3, 2013.
71. D. Klemm, H. P. Schmauder, T. Heinze, in *Biopolymers: Polysaccharides II*, ed. S. De Baets, E. J. Vandamme and A. Steinbuchel, Wiley-VCH, Weinheim, 2002, p. 275.
72. R. F. Tester, J. Karkalas, in *Biopolymers: Polysaccharides II*, ed. S. De Baets, E. J. Vandamme and A. Steinbuchel, Wiley-VCH, Weinheim, 2002, p. 381.
73. D. Lourdin, L. Coignard, H. Bizot and P. Colonna, *Polymer*, 1997, **38**, 5401–5406.
74. M. Avella, J. J. de Vlieger, M. E. Enrico, S. Fischer, P. Vacca and M. G. Volpe, *Food Chem.*, 2005, **93**, 467–474.
75. L. Averous, C. Fringant and L. Moro, *Starch*, 2001, **53**, 368–371.
76. Y. Habibi, A. L. Goffin, N. Schiltz, E. Duquesne, P. Dubois and A. Dufresne, *J. Mater. Chem.*, 2008, **18**, 5002–5010.
77. J. Hafren and A. Cordova, *Macromol. Rapid Commun.*, 2005, **26**, 82–86.
78. H. Lonnberg, Q. Zhou, H. Brumer, T. T. Teeri, E. Malmstrom and A. Hult, *Biomacromolecules*, 2006, **7**, 2178–2185.
79. N. Berezina, J. Nys and B. Yada, *Chem. Eng. Trans.*, 2013, **32**, 1003–1008.
80. A. Franck, L. De Leenheer, in *Biopolymers: Polysaccharides II*, ed. S. De Baets, E. J. Vandamme and A. Steinbuchel, Wiley-VCH, Weinheim, 2002, p. 439.
81. B. Kleesem, B. Sykura and H. J. Zunft, *Am. J. Clin. Nutr.*, 1997, **65**, 1397–1402.
82. I. R. Rowland, C. J. Rumney, J. T. Coutts and L. C. Lievense, *Carcinogenesis*, 1998, **19**, 281–285.
83. M. C. Ralet, E. Bonnin, J. F. Thibault, in *Biopolymers: Polysaccharides II*, ed. S. De Baets, E. J. Vandamme and A. Steinbuchel, Wiley-VCH, Weinheim, 2002, p. 345.
84. D. Mohnen, *Curr. Opin. Plant Biol.*, 2008, **11**, 266–277.
85. R. K. Portenoy, A. W. Burton, N. Gabrail and D. Taylor, *Pain*, 2010, **151**, 617–624.
86. M. C. Peter, in *Biopolymers: Polysaccharides II*, ed. S. De Baets, E. J. Vandamme and A. Steinbuchel, Wiley-VCH, Weinheim, 2002, p. 481.
87. P. Pochanavanich and W. Suntornsuk, *Lett. Appl. Microbiol.*, 2002, **35**, 17–21.
88. M. N. V. R. Kumar, *Reactive Funct. Polymers*, 2000, **46**, 1–27.
89. E. Valepyn, N. Berezina and M. Paquot, *Adv. Microbiol.*, 2012, **2**, 488–496.

90. J. Preiss, in *Biopolymers: Polysaccharides I*, ed. S. De Baets, E. J. Vandamme and A. Steinbuchel, Wiley-VCH, Weinheim, 2002, p. 21.

91. C. Heidrich, W. Vollmer, in *Biopolymers: Polysaccharides I*, ed. S. De Baets, E. J. Vandamme and A. Steinbuchel, Wiley-VCH, Weinheim, 2002, p. 431.

92. V. Lazarevic, H. M. Pooley, C. Mauel, D. Karamata, in *Biopolymers: Polysaccharides I*, ed. S. De Baets, E. J. Vandamme and A. Steinbuchel, Wiley-VCH, Weinheim, 2002, p. 465.

93. J. Courtois, *Curr. Opin. Microbiol.*, 2009, **12**, 261–273.

94. T. Kuda, T. Yano, N. Matsuda and M. Nishizawa, *Food Chem.*, 2005, **91**, 745–749.

95. M. E. El-Boshy, A. M. El-Ashram, F. M. AbdelHamid and H. A. Gadalla, *Fish Shellfish Immun.*, 2010, **28**, 802–808.

96. U. Rau, in *Biopolymers: Polysaccharides II*, ed. S. De Baets, E. J. Vandamme and A. Steinbuchel, Wiley-VCH, Weinheim, 2002, p. 61.

97. S. Bielecki, A. krystynowicz, M. Turkiewicz, H. Kalinowska, in *Biopolymers: Polysaccharides I*, ed. S. De Baets, E. J. Vandamme and A. Steinbuchel, Wiley-VCH, Weinheim, 2002, p. 37.

98. T. D. Leathers, in *Biopolymers: Polysaccharides I*, ed. S. De Baets, E. J. Vandamme and A. Steinbuchel, Wiley-VCH, Weinheim, 2002, p. 299.

99. K. Born, V. Langendorff, P. Boulenguer, in *Biopolymers: Polysaccharides I*, ed. S. De Baets, E. J. Vandamme and A. Steinbuchel, Wiley-VCH, Weinheim, 2002, p. 259.

100. T. J. Pollock, in *Biopolymers: Polysaccharides I*, ed. S. De Baets, E. J. Vandamme and A. Steinbuchel, Wiley-VCH, Weinheim, 2002, p. 239.

101. I. Giavasis, L. M. Harvey, B. McNeil, in *Biopolymers: Polysaccharides II*, ed. S. De Baets, E. J. Vandamme and A. Steinbuchel, Wiley-VCH, Weinheim, 2002, p. 37.

102. I. Y. Lee, in *Biopolymers: Polysaccharides I*, ed. S. De Baets, E. J. Vandamme and A. Steinbuchel, Wiley-VCH, Weinheim, 2002, p. 135.

103. B. H. A. Rehm, in *Biopolymers: Polysaccharides I*, ed. S. De Baets, E. J. Vandamme and A. Steinbuchel, Wiley-VCH, Weinheim, 2002, p. 179.

104. T. D. Leathers, in *Biopolymers: Polysaccharides II*, ed. S. De Baets, E. J. Vandamme and A. Steinbuchel, Wiley-VCH, Weinheim, 2002, p. 1.

105. K. I. Draget, O. Smidsrod, G. Skjak-Braek, in *Biopolymers: Polysaccharides II*, ed. S. De Baets, E. J. Vandamme and A. Steinbuchel, Wiley-VCH, Weinheim, 2002, p. 215.

106. H. von Döhren in *Biopolymers: Polyamides and Complex Proteinaceous Materials I*, ed. S. R. Fahnestock and A. Steinbüchel, Wiley-VCH, Weinheim, 2003, p 51.

107. H. Carillo-Navas, J. Cruz-Oliveras, V. Varela-Guerrero, L. Alamilla-Beltran, E. J. Vernon-Carter and C. Perez-Alonso, *Carb. Polym.*, 2012, **87**, 1231.

108. A. S. Hoffman and P. S. Stayton, *Prog. Polym. Sci.*, 2007, **32**, 922.

109. T. Yoshida, J. Hiraki and T. Nagasawa in *Biopolymers: Polyamides and Complex Proteinaceous Materials I*, ed. S. R. Fahnestock and A. Steinbüchel, Wiley-VCH, Weinheim, 2003, p. 107.
110. W. Joengten, N. Müller, A. Mitschker and H. Scmidt in *Biopolymers: Polyamides and Complex Proteinaceous Materials I*, ed. S. R. Fahnestock and A. Steinbüchel, Wiley-VCH, Weinheim, 2003, p. 175.
111. J.-M. Raquez, M. Deleglise, M.-F. Lacrampe and P. Krawczak, *Prog. Polym. Sci.*, 2010, **35**, 487.
112. A. Sionkowa, *Prog. Polym. Sci.*, 2011, **36**, 1254.
113. I. Tchmutin, N. Ryvkina, N. Saha and P. Saha, *Polym. Degrad. Stab.*, 2004, **86**, 411.
114. A. K. Mohanty, W. Liu, P. Tummula, L. T. Drzal, M. Manjusri and R. Narayan, in *Natural Fibers, Biopolymers, and Biocomposites*, ed. A. K. Mohanty, M. Misra and L. T. Drzal, Taylor & Francis, Boca Raton, 2005, p. 699.
115. F. Chen and J. Zhang, *Polymers*, 2009, **50**, 3770.
116. L. Yu, K. Dean and L. Li, *Prog. Polym. Sci.*, 2006, **31**, 576.
117. B. Monties and K. Fukushima, in *Biopolymers: Lignin, Humic Substances and Coal*, ed. M. Hofrichter and A. Steinbüchel, Wiley-VCH, Weinheim, 2001, p. 1.
118. G. Brunow, in *Biopolymers: Lignin, Humic Substances and Coal*, ed. M. Hofrichter and A. Steinbüchel, Wiley-VCH, Weinheim, 2001, p. 65.
119. M. N. Mohamad Ibrahim, H. Azian and M. R. Mohd Yusop, *J. Technol.*, 2006, **44**, 83–94.
120. A. S. Jonsson, A. K. Nordin and O. Wallberg, *Chem. Eng. Res. Design*, 2008, **86**, 1271–1280.
121. V. M. Roberts, V. Stein, T. Reiner, A. Lemonidou, X. Li and J. A. Lercher, *Chem. Eur. J*, 2011, **17**, 5939–5948.
122. J. Li, G. Henriksson and G. Gellerstedt, *Bioresour. Technol*, 2007, **98**, 3061–3068.
123. M. P. Pandey and C. S. Kim, *Chem. Eng. Technol.*, 2011, **34**, 29–41.
124. D. Stewart, *Ind. Crops Prod*, 2008, **27**, 202–207.
125. E. Svensson and T. Berntsson, *Appl., Therm. Eng.*, DOI: 10.1016/j.applthermaleng.2009.02.010.
126. A. S. Jonsson and O. Walberg, *Desalination*, 2009, **237**, 254–267.
127. Y. J. Wee, J. N. Kim and H. W. Ryu, *Food Technol. Biotechnol.*, 2006, **44**, 163–172.
128. M. Sauer, D. Porro, D. Mattanovich and P. Branduardi, *Trends Biotechnol.*, 2008, **26**, 100–108.
129. J. Lunt, *Polym. Degrad. Stab.*, 1998, **59**, 142–152.
130. D. Carlotta, *J. Polym. Environ.*, 2002, **9**, 63–84.
131. H. Tsuji and Y. Ikada, *Polymer*, 1999, **40**, 6699–6708.
132. S. Okino, M. Suda, K. Fujikura, M. Inui and H. Yukawa, *Appl. Microiol. Biotechnol.*, 2008, **78**, 449–454.
133. K. Okamoto, K. Toshima and S. Matsumura, *Macromol. Biosci.*, 2005, **5**, 813–820.

134. T. Tsukegi, T. Motoyama, Y. Shirai, H. Nishida and T. Endo, *Polym. Degrad. Stab.*, 2007, **92**, 552–559.

135. T. Motoyama, T. Tsukegi, Y. Shitai, H. Nishida and T. Endo, *Polym. Degrad. Stab.*, 2007, **92**, 1350–1358.

136. C. Gao, J. Qiu, J. Li, C. Ma, H. Tang and P. Xu, *Bioresour. Technol.*, 2009, **100**, 1878–1880.

137. C. Ma, J. Gao, J. Qiu, W. Hao, A. Liu, Y. Wang, M. Zhang and P. Wang, *Appl. Microbiol. Biotechnol.*, 2007, 77, 91.

138. B. Martin-Matute, J. E. Backvall, in *Organic Synthesis with Enzymes in Non-Aqueous Media*, Wiley-VCH, Weinheim, 2008, pp. 113–114.

139. C. Inaba, K. Maekawa, H. Morisaka, K. Kuroda and M. Ueda, *Appl. Microbiol. Biotechnol.*, 2009, **83**, 859–864.

140. N. Berezina, N. Landercy, P. A. Mariage and B. Moreau, *J. Org. Chem.*, 2013, **1**, 20–23.

141. J. R. Dorgan, H. Lehermeier and M. Mang, *J. Polym. Environ.*, 2000, **8**, 1–9.

142. N. Nagasawa, A. Ayako, S. Kanasawa, T. Yagi, H. Mitomo, F. Yoshii and M. Tamada, *Nucl. Instr. Meth. Phys. Res. B*, 2005, **236**, 611–616.

143. F. Yang, R. Murugan, S. Wang and S. Ramakrishna, *Biomaterials*, 2005, **26**, 2603–2610.

144. T. Homan, C. Tag, H. Biebl, W. D. Deckwer and B. Schink, *Appl. Microbiol. Biotechnol.*, 1990, **33**, 121–126.

145. A. Drozdzynska, K. Leja and K. Czaczyk, *J. Biotechnol. Comput. Biol. Bionanotechnol.*, 2011, **92**, 92–100.

146. H. Ahmed, O. Naouel and B. Djilali, *Biotechnol. Biomaterials*, 2012, **2**, 134.

147. A. A. Gatenby, *et al.*, 1998, WO9821339.

148. V. Nagarajan, C. E. Nakamura, 1998, US5821092.

149. C. E. Nakamura, A. A. Gatenby, 2000, US6013494.

150. D. R. Kelsey, 1998, US5705575.

151. J. Kaskey, G. Sulugiuc, *Bloomberg.net*, 2011, January 11.

152. M. V. Guettler, M. K. Jain, 1996, US5521075.

153. E. Delwiche, *J. Bacteriol.*, 1950, **59**, 439–442.

154. L. Stols and M. I. Donnelly, *Appl. Environ. Microbiol.*, 1997, **63**, 2695–2701.

155. J. W. Lee, H. U. Kim, S. Choi, J. Yi and S. Y. Lee, *Curr. Opin. Biotechnol.*, 2011, **22**, 758–767.

156. N. Jacquel, *et al.*, *J. Polym. Sci.*, 2011, **49**, 5301–5312.

157. P. Rizzarelli and S. Carroccio, *Polym. Degrad. Stab.*, 2009, **94**, 1825–1838.

158. J. Xu and B. H. Guo, *Biotechnol. J*, 2010, **5**, 1149–1163.

159. S. Sugihara, K. Toshima and S. Matsumura, *Macromol. Rapid Commun.*, 2006, **27**, 203–207.

160. A. Kondo, S. Sugihara, M. Kuwahara, K. Toshima and S. Matsumura, *Macromol. Biosci.*, 2008, **8**, 533–539.

161. A. Kondo, S. Sugihara, K. Okamoto, Y. Tsuneizumi, K. Toshima and S. Matsumura, *Polym. Biocat. Biomat.*, 2008, 246–262.

162. P. Wells and C. Zapata, *J. Indus. Ecol.*, 2012, **16**, 665–668.
163. P. J. Halley and J. R. Dorgan, *MRS Bull.*, 2011, **36**, 687–691.
164. J. Gotro, *Polymer Innovation Blog*, 2013, March 11.
165. A. C. Trefzer, S. C. H. J. Turk, 2012, WO 2012/031910 A2.
166. P. C. Raemakers-Franken *et al.*, 2009, US 2009/0137759 A1.
167. W. Buijs, H. F. W. Wolkers, R. P. M. Guit, F. P. W. Agterberg, 2001, US 6.194.572 B1.
168. T. Polen, M. Spelberg, M. Bott, *J. Biotechnol.*, 2012, http://dx.doi.org/10.1016/j.jbioyec.2012.07.008.
169. K. A. Curran, J. M. Leavitt, A. S. Karim and H. S. Alper, *Metabol. Eng.*, 2013, **15**, 55–66.
170. A. M. Warhurst, K. F. Clarke, R. A. Hill, R. A. Holt and C. A. Fewson, *Biotechnol. Lett.*, 1994, **16**, 513–516.
171. D. Guzman, *ICIS Green Chemicals*, 2012, June 6.
172. Q. Fing and M. A. Hanna, *Cereal. Chem.*, 2000, 77, 779–783.
173. C. Li, J. Zhang, Y. Li, S. Moran, G. Khang and Z. Ge, *Biomed. Mater.*, 2013, **8**, 1–13.
174. A. Ochi, K. Matsumoto, T. Ooba, K. Sakai, T. Tsuge and S. Taguchi, *Appl. Microbiol. Biotechnol.*, 2013, **97**, 3441–3447.
175. J. H. Shut, *Plastics Technol.*, 2008, November 66–69.
176. R. Harracksingh. *Isis.com*, 2012, June 29.
177. D. Guzman, *Green Chemicals Blog*, November 11, 2012.
178. Bioplastic Innovation website, July 12, 2011.
179. D. Guzman, *Green Chemicals Blog*, March 12, 2013.
180. D. Guzman, *Green Chemicals Blog*, March 18, 2013.
181. Bioplastic Innovation website, March 23, 2012.
182. P. Prehm, in *Biopolymers: Polysaccharides I*, ed. S. De Baets, E. J. Vandamme and A. Steinbuchel, Wiley-VCH, Weinheim, 2002, p. 379.
183. M. Stredansky, in *Biopolymers: Polysaccharides I*, ed. S. De Baets, E. J. Vandamme and A. Steinbuchel, Wiley-VCH, Weinheim, 2002, p. 159.
184. J. Courtois, B. Courtois, in *Biopolymers: Polysaccharides I*, ed. S. De Baets, E. J. Vandamme and A. Steinbuchel, Wiley-VCH, Weinheim, 2002, p. 213.
185. G. L. Cote, in *Biopolymers: Polysaccharides I*, ed. S. De Baets, E. J. Vandamme and A. Steinbuchel, Wiley-VCH, Weinheim, 2002, p. 323.
186. S. K. Rhee, K. B. Song, C. H. Kim, B. S. Park, E. K. Jang, K. H. Jang, in *Biopolymers: Polysaccharides I*, ed. S. De Baets, E. J. Vandamme and A. Steinbuchel, Wiley-VCH, Weinheim, 2002, p. 351.
187. F. Van de Velde, G. A. De Ruiter, in *Biopolymers: Polysaccharides II*, ed. S. De Baets, E. J. Vandamme and A. Steinbuchel, Wiley-VCH, Weinheim, 2002, p. 245.
188. G. Paradossi, F. Cavalieri and E. Chiessi, *Macromolecules*, 2002, **35**, 6404–6411.

CHAPTER 2

Fundamentals and Biotechnological Applications of Downstream Processing Technologies

RAQUEL MANOZZO GALANTE,[a]
GUSTAVO GRACIANO FONSECA,[a] NATHALIE BEREZINA,[b]
THIAGO CAON,[c] FARAYDE MATTA FAKHOURI[a] AND
SILVIA MARIA MARTELLI*[a]

[a] Faculty of Engineering, Federal University of Grande Dourados, Dourados, Brazil; [b] Materia Nova, Rue des Foudriers 1, 7822 Ghislenghien, Belgium; [c] Department of Pharmaceutical Sciences, Federal University of Santa Catarina, Brazil
*Email: smmartelli@gmail.com

2.1 Introduction

Industrial biotechnology comprises sustainable processing and production of food, chemicals, materials and fuels.[1,2] Our global challenges require sustainable value creation for science, industry, society and the environment. In this sense, biotechnological processes through the use of enzymes and/or microorganisms may be used, manufacturing innovative bioproducts, reducing waste and/or using disposable materials in order to add value to the products.[3–7] The biotechnology industry has already provided many benefits for a long time, but with the advance of new technologies and

RSC Green Chemistry No. 27
Renewable Resources for Biorefineries
Edited by Carol Sze Ki Lin and Rafael Luque

a much deeper understanding of cell metabolism and materials science, new opportunities have been identified in the last few years.[8] The products are immersed in a broad range of industrial sectors including chemicals, pharmaceuticals, food and feed, detergents, paper and pulp, textiles, energy, materials and polymers.[9–12]

Despite the concept of the biotechnological process that are discussed in this book, it is worthwhile remembering all the parts comprising the process of obtaining a product through a biotechnological process. Overall, the process can be divided into two parts: the upstream processing (USP) and the downstream processing (DSP). All the initial steps are included in the upstream processing, which include the selection of microorganism (microorganisms screening, suitable strain selection, and if necessary, organism genetic modification), the formulation and optimization of the medium (proper growth conditions), the small-scale bioreactor production (batch, fed-batch, continuous), process control parameters. The upstream stages have been extensively studied and significantly improved over time.[13,14] When this part is successfully achieved, the next step is the DSP, which involves biomass/product separation and the subsequent product purification, concentration and, finally, its sale. It is important to emphasize that the correct selection made during the USP (cell lines, fermentation parameters and so on) can directly impact on the cost of the DSP, which will reflect on the process suitability.

From the beginning of the 1990s, much progress has been made in biological upstream processing, with productive high-expression systems, effective clone selection, defined culture media, along with improvement in analytical methods, which were developed to characterize and quantify the target products. These advances were extremely important because they helped to characterize the starting materials, providing more insight into quality and mass balance during the process as well helping greatly in improving product yield. As a consequence of these developments in the upstream processing, new challenges in downstream processing were generated. Additional challenges for downstream processing include treating high upstream quantities in a short time, integrating new technologies, purifying molecules in a contained system that enables the handling of the product, and, of course, ensuring that processes are cost effective.[15]

The fermentation processes (upstream processing) used to produce industrial biotechnological products present unique separation and purification challenges. Fermentation systems can contain several compounds that can interfere with downstream separation processes, making it costly and time demanding.[16] Among the medium-specific characteristics, the most common ones are the high amount of water, the presence of organic and inorganic molecules from the fermentation broth, as well as extracellular and intracellular metabolites from dead cells and cell fragments. There are no general purification processes for all products. The main steps in downstream processing can be divided into cellular inactivation, clarification and disruption, product extraction (recovery) and purification. The number and type of the steps will depend on material and purity level demanded.

Traditional methods to purify biomolecules involve sequential stages, such as dialysis, ionic and affinity chromatography, among others. Liquid/ liquid extractions (LLE) have also been an interesting alternative to purify biotechnological products because several features of the early processing steps can be combined into a single operation.[17]

It is worth noting that in spite of several steps comprising the downstream process being well defined, each step can be formed by several unit operations. For instance, after precipitation by adding a salt, a dialysis step is necessary to adjust the ionic force, followed (or not) by a chromatography step. Since the separation steps involved at the end of the production step are highly costly and can reduce the yield of the overall process, well-designed experiments require large efforts to minimize the number of steps necessary to recover the desired product.[18–20]

The selection of unit operations to be used in a specific recovery process is linked directly to the type of material to be obtained, its physicochemical characteristics, level of impurities and the ultimate use of the product (ranging from animal feed to human parenteral).[21,22] Obviously, products that might be used in therapeutics are the ones which require higher levels of purity and therefore the complexity of the purification process is greater.[23–26] The main objective of this chapter is to describe the basic characteristics of the most commonly used downstream processing steps in industrial biotechnology after the fermentation process and biocatalysis, presenting their advantages and disadvantages as well as their main applications.

2.2 Downstream Processing Steps

A minimal number of steps, with each step being simple, is highly desired during DSP processing. These steps are usually defined by regarding the physicochemical properties of the product and also the final application of the material, yield, costs and level of purity. A bioseparation process must combine high selectivity (or resolution) with high throughput (productivity). To achieve these goals, a strategy has been developed, involving use of low-resolution techniques (primary recovery) first for recovery and isolation followed by high-resolution techniques for purification and polishing.[27] The general DSP steps used for the purification of biomolecules are shown in Figure 2.1.

The conventional sequence of separation starts with the use of techniques that separate components having large differences in their physicochemical properties and ends with separation of molecules whose properties are similar.[28] Solid/liquid separation can be achieved through operations such as filtration, sedimentation, centrifugation (sedimentation using the gravitational force), flocculation and flotation. Extracellular products (if any) may be rapidly extracted since they may be easily found in the liquid phase. On the other hand, the extraction of intracellular products requires an additional step – cell disruption – which is usually accomplished by mechanical (maceration, bead mill, homogenization, ultrasonication, freezing and

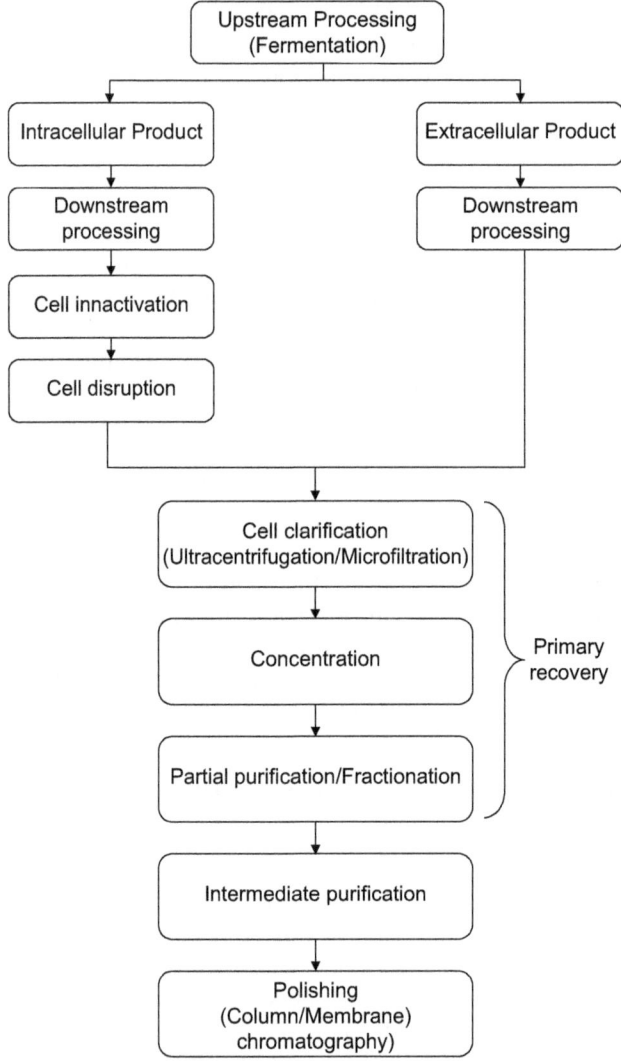

Figure 2.1   Flow chart for the downstream processing of a general biotechnological product.
Adapted from Jahanshahi and Najafpour (2007) and Prasad (2010).[29,34]

griding) or non-mechanical processes (drying, heat shock, osmotic shock, freeze thaw, organic solvents, chaotropic agents, alkali agents, detergents or enzymes). The extraction step can be performed by using liquid/liquid extraction, whole broth extraction, aqueous multi-phase extraction or supercritical fluid. After the extraction, the desired material can be concentrated by techniques such as evaporation, membrane filtration, adsorption, ion-exchange resins or precipitation. The last step is the polishing, which may be performed using crystallization, freeze-drying, spray drying or sterile filtration.[29]

As mentioned before, the basis of bioseparation and unit operation is based on the differences of physicochemical properties of the materials (Figure 2.2). Chromatography methods are not always the best option due to variable yield losses and high costs.[30] Therefore, various attempts have been made to find new separation processes focusing on cost reduction. Among these options, a versatile and promising technique is that of the foam fractionation (Figure 2.2), an adsorptive bubble separation technique in which the principle of separation is based on the differences in the surface activity of molecules. It has been used to separate proteins, but it can also be used for other purposes (*e.g.* the concentration of plant secondary metabolites).[31]

Foam is a gas/liquid dispersion system, with gas bubbles forming an inner non-continuous phase and liquid forming a continuous phase. As bubbles pass through a liquid solution, surface-active compounds preferentially adsorb onto a bubble surface.[32] The surface-active compounds can be carried out from the liquid phase by these bubbles into a foam phase, which can be formed when these bubbles accumulate above the gas/liquid pool interface. The most strongly surface active component or component with the largest bubble net adsorption rate in the liquid solution will have the highest relative adsorption in the foam phase. When the foam phase collapses to form a new liquid phase, a liquid solution can be produced with a

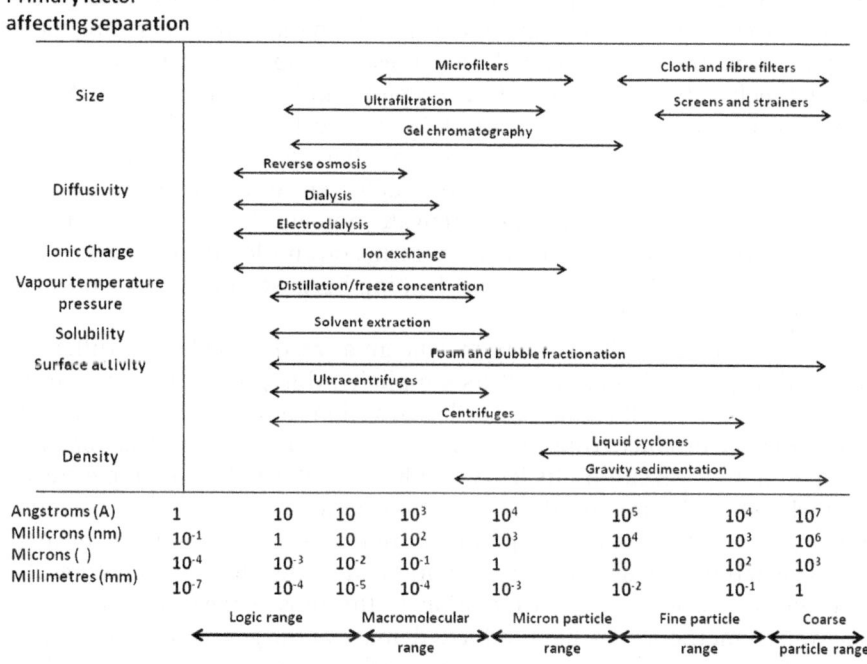

Figure 2.2 Basis of bioseparation and the unit operation involved.

solute concentration several times higher than the original solution due to the large prior bubble surface area and small liquid content of the foam.[33]

2.2.1   Cell Inactivation

In the inactivation process, microorganisms are completely inactivated but at the same time retain the original cell characteristics (for example, immunogenic sites of microorganisms should be retained) as much as possible. In some situations, cell inactivation is performed to prevent the consumption of desired product by microorganisms. Polyhydroxyalkanoates, for example, can be produced and stored as intracellular granules by microorganisms under specific conditions. If microorganisms are not inactivated after fermentation, these compounds can be easily metabolized.[34,35]

This step is also highly relevant in the manufacture of vaccines. Inactivated vaccines consist of virus particles which are grown in culture and then killed.[36–38] Cell inactivation should be able to preserve the immunogenicity or epitopes associated with protection. It is well known that some reagents used to inactivate cells can change the outer membrane antigens with the risk of a reduced immunogenicity of the vaccine, thus leading to loss of efficacy.[39] To achieve inactivation, it is possible to use chemical (acids, alkalis, alcohols, surface active agents, phenols, oxidizing agents, formaldehyde, monochloramine, *etc.*) or physical agents (heat, UV irradiation and high pressure). These last method usually reduce the number of purification steps and toxicity issues, thus reducing the production cost.[40] Fundamental mechanisms by which viruses are inactivated have not been fully characterized regarding their complex structure.[41–43] In general, they present a lipid envelope, structural proteins such as the capsid or surface receptors, and the nucleic acid (DNA or RNA), which can all be targets of attack by chemical biocides.[41]

Each method of inactivation has special characteristics that need to be taken into account. For example, solvent/detergent is very effective against enveloped viruses, but does not inactivate non-enveloped viruses. If the hepatitis B virus (HBV) is a principal concern, solvent/detergent may have an advantage over methods that employ heating because HBV is known to be relatively heat stable.[44]

Although the thermal treatment may be an attractive method in different inactivation protocols because of its simplicity, this procedure can significantly reduce the efficiency of cell disruption during homogenization depending on product to be processed.[45] In the processing of *E. coli*, for example, homogenization efficiency is adversely affected by a simple treatment whereby cells are raised to 65 °C from stationary phase of the bacteria growth curve. Protein-based products may also be degraded during homogenization because of heat generation since they may be denatured.[45]

Currently, ultraviolet (UV) irradiation is the most widely used physical inactivation method.[46] UV light induces damage at the genomic level of cells, mainly due to the fact that absorption is much stronger in DNA compared to proteins and other biological molecules.[47]

Among chemical methods, monochloramine and free chlorine are used extensively to inactivate viruses.[48,49] Inactivation can also occur by degradation or disruption of any one of the viruses structures or by all of them, depending on the chemistry of the disinfectant.[41]

In summary, it is worth noting that the selection of an inactivation procedure will depend on the type/complexity of the material to be obtained (for instance, enveloped viruses are more difficult to inactivate than those non enveloped), as well as its compatibility with other processing steps. If the final product is not consumed during the processing steps, spontaneous inactivation due to mechanical stress or other phenomena could also be observed (for example, filtration processes involving change of phase – liquid to solid – may be traumatic for the virus particle). In this situation, an individual inactivation step would not be necessary, resulting in cost reductions.

2.2.2 Clarification

The clarification process is the unit operation designed to separate the cells from the liquid broth immediately after the fermentation. The most commonly used techniques are filtration and centrifugation.[50,51] When the product is inside the cells, cell disruption is necessary after the clarification. Intracellular products make the purification process of biomolecules more complex compared to extracellular compounds. After the cell disruption, there is an increase in viscosity of the medium due to the release of nucleotides and the desired compound is delivered together with all other molecules, requiring additional steps for its purification.[52]

Clarification processes are normally used to separate the suspended cells from a fluid stream primarily by direct interception and sieving, where the particles are caught in or on the surface of the filter medium. In addition, some liquid filters' removal capabilities can be achieved or enhanced by charge effects, in which case they may retain some molecules by adsorption.[21,53]

The terminology 'suspended cells' is used because the microbial cells may be considered in fact to be suspended solids. It is possible to separate them from liquid media using operations such as filtration and centrifugation as these processes were developed to mechanically separate fluid and particulate phases.[54] At the laboratory scale, removal of cells and large cell debris is achieved by low-speed centrifugation and microfiltration. Membrane fouling is the main problem faced during filtration since it reduces the flow rate over time. To keep the process time within reasonable limits, without increasing operating pressures that may affect the stability of the material to be isolated, it is often necessary to restrict the volume to be filtered. The introduction of a centrifugation step before membrane filtration also avoids membrane clogging, particularly when the volume of supernatant to be passed per filter is high. It is also convenient to filter crude supernatants through a series of membranes with decreasing pore size to minimize membrane clogging.[55]

Electro-coagulation has also been proposed in the clarification step of plant-based bioprocesses. It involves the generation of coagulants by the electrochemical dissolution of metal ions from the anode with simultaneous formation of hydroxyl ions and hydrogen gas at the cathode. The metal ions form the flocculate, which traps the contaminants, while hydrogen helps to float the flocculated particles.[54] This technique is relatively effective and easy to operate. It can also be used as a protein pre-purification step since it is able to select different proteins based on their isoelectric point (pI.).[56]

Traditionally, centrifugation followed by a combination of filtration techniques (tangential-flow filtration or depth filtration) are used for clarifying complex cell culture broths.[57] The main aspects of filtration and centrifugation are briefly described below.

2.2.2.1   Filtration

The filtration of fermentation media, purification buffers, and protein product pools is standard practice in industry. Microfiltration is used extensively for medium exchange and harvest. Ultrafiltration can be found in virtually every biotechnology process since it allows gentle processing of large volumes of supernatant in a relatively short time. While microfiltration is used to retain suspended particles presenting size larger than 0.10–5 μm, ultrafiltration is able to retain macromolecules of approximately 0.001–0.02 μm. A considerable number of mammalian cell processes have used filtration as an integral part of the overall approach for viral clearance. Depth filters have also been extensively used for the clarification of both mammalian and bacterial feed streams.[58]

The filter process consists of the mechanical separation of solid particles from a liquid suspension by using a porous medium. When the suspension is forced through the bed, the solid becomes trapped over the *filter medium* forming a deposit, the *filter cake*, which has its thickness increased during the operation. The liquid that passes through the bed is the *filtrate*.[59] The filtration process mentioned above is described schematically in Figure 2.3.

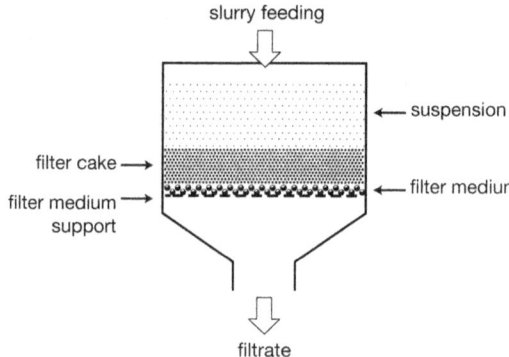

Figure 2.3   Operational principle of a conventional filter.

Conventional filtration is used to separate large volumes of diluted cells suspensions (>1 L), extracellular products and mould suspensions because the mycelium has very low density and it is difficult to separate the fluid by centrifugation.[59]

In the process of filtration, the medium resistance over the fluid flow increases with the formation of a cake or with time as the filter medium is obstructed. The main variables of interest are the flow velocity through the filter and the pressure drop in the unit.

Darcy's law describes the filtration processes, correlating variables of interest according to equation 2.1. The velocity of the fluid that passes through the filter medium can be determined by equation 2.2.[59,60]

$$v = \frac{k \, \Delta P}{\mu} \tag{2.1}$$

$$v = \frac{1}{A} \frac{dV}{dt} \tag{2.2}$$

where $v =$ flow velocity (m s^{-1}); $k =$ bed permeability (m^2); $\Delta P =$ pressure difference across the bed (N m^{-2}); $\mu =$ viscosity (kg m^{-1} s^{-1}); $1/k =$ filter bed resistance; $A =$ filtration area (m^2); $V =$ volume of filtrate (m^3); $t =$ filtration time (s).

The filtration resistance (equation 2.3) can be separated in two parts: one corresponding to the filter medium resistance (R_m), and another related to the filter cake (R_c). The second part depends on the particles' mass fraction in suspension, X (kg m^{-3}), the filtrate volume (V), the filtration area (A) and the specific resistance of the cake, α (m kg^{-1}), according to equation 2.4:

$$\frac{1}{k} = R_m + R_c \tag{2.3}$$

$$R_c = \alpha X \frac{V}{A} \tag{2.4}$$

The specific resistance value (α) in the case of incompressible cakes assumes a constant value. For compressible cakes, the resistance varies with the pressure according to equation 2.5:

$$\alpha = \alpha_0 (\Delta P)^s \tag{2.5}$$

where: $\alpha_0 =$ empirical constant; $s =$ compressibility constant, which is zero for compressible cakes.

Finally, equation 2.6 is obtained combining equations 2.1 to 2.4, which correlates all variables of the filtration process.

$$\frac{1}{A} \frac{dV}{dt} = \frac{\Delta P}{\mu R_m + \mu \alpha X \frac{V}{A}} \tag{2.6}$$

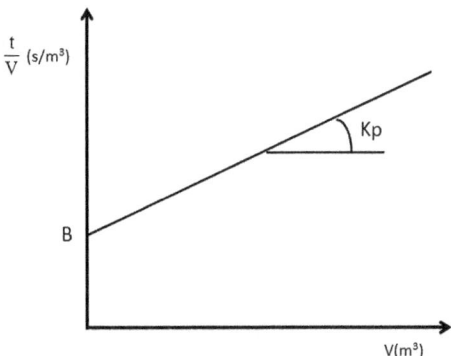

Figure 2.4 Linearization of the experimental data in the filtration process.

When integrating equation 2.6, the filtrate volume is considered zero at time $t = 0$ and V at any instant t. This results in equation 2.7, which relates the collected filtrate volume at any time t:

$$\frac{t}{V} = \frac{\mu \alpha X}{2A^2 \Delta P} V + \frac{\mu R_{\mathrm{m}}}{A \Delta P} = \frac{K_{\mathrm{p}}}{2} V + B \tag{2.7}$$

where: $K_{\mathrm{p}} = \dfrac{\mu \alpha X}{A^2 \Delta P} (\mathrm{s\ m}^{-6}); B = \dfrac{\mu R_{\mathrm{m}}}{A \Delta P} (\mathrm{s\ m}^3).$

It is possible to determine experimentally the volume of filtrate by filtration time, which can be calculated from the analysis of t/V versus V, as shown in Figure 2.4. Regarding the function obtained, it is possible to calculate the constants α and R_{m}.[59,60]

2.2.2.2 *Types of Equipment*

Among the traditional types of filters, the use of a plate and frame filter is widespread because it results in more dehydrated cakes compared with other groups. On the other hand, this type of filter is difficult to handle, restricting its application to small volumes.[59]

For separation of large volumes of suspension, vacuum rotating filters are recommended (Figure 2.5). The operation of the rotating vacuum filter is characterized by producing dried cakes of small thickness (less than 1 cm). Furthermore, the cake is easily removed and the operating parameters are easy to control.

The filtration is carried out on the filtering medium that covers the equipment's cylindrical surface. While the filter rotates, the following operations occur: fluid vacuum drainage from the cake; cake washing with the aid of a shower; new fluid drainage; and removal of the cake that is in contact with the filter medium.[62]

Among the disadvantages of the rotating vacuum filter are the high energy consumption due to the vacuum pump and the constant washing required to the filter media.

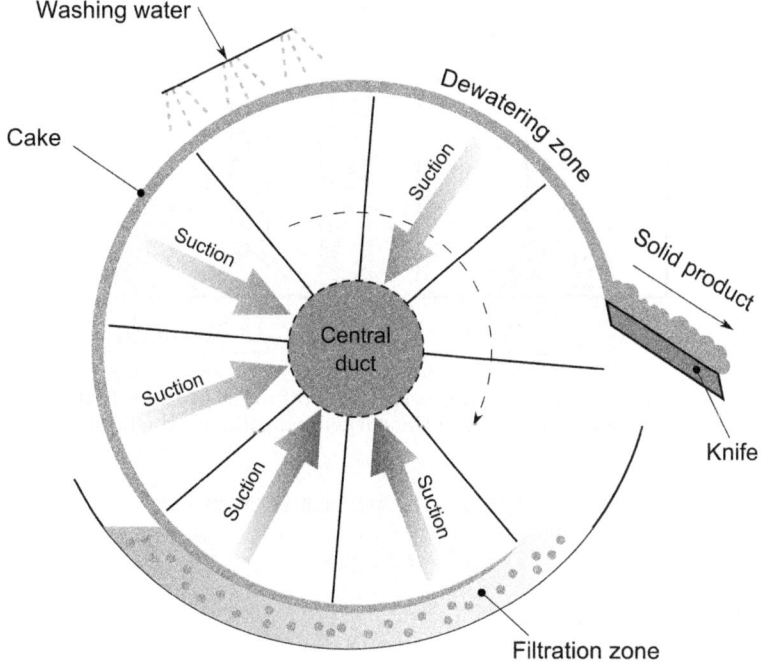

Figure 2.5 Schematic diagram of a rotating vacuum filter, modified from Geankoplis (2003).[61]

2.2.2.3 Centrifugation

Centrifugation is a mechanical separation technique, which accelerates the phenomenon of sedimentation through the imposition of rotational motion in a liquid/particle suspension.[62] In the centrifugation equipment, the centrifugal force causes the particles to move radially away from the axis of rotation, as shown in Figure 2.6.

In the centrifugal separation, the centrifugal acceleration in a circular motion can be represented by equation 2.8:[61]

$$a = r\omega^2 \qquad (2.8)$$

where a = centrifugal acceleration (m s^{-2}); r = radial distance from the centre of rotation (m), ω = angular velocity (rad s^{-1}).

Centrifugation is based on the density difference between the particle (ρ_p) and the liquid medium (ρ), the viscosity of the liquid medium (μ), the driving force and the particle diameter (D_p), according to equation 2.9, where v_p represents the particle velocity in the centrifugal field (m s^{-1}).[59]

$$v_p = \frac{\omega^2 r D_p^2 (\rho_p - \rho)}{18\mu} \qquad (2.9)$$

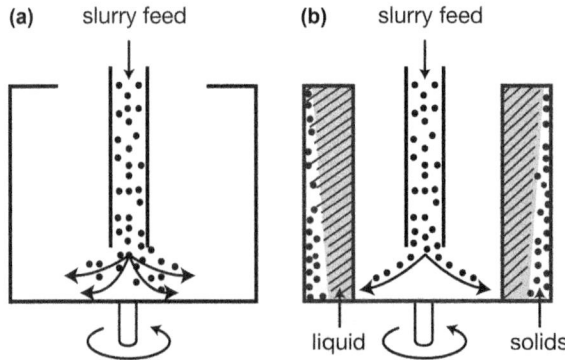

Figure 2.6   Diagram of centrifugal separation: (a) input of the initial suspension and (b) precipitation of the suspended solids in the liquid.[61]

The centrifugal force acting on the particle is given by:

$$F_c = ma = mr\omega^2 \qquad (2.10)$$

where F_c = centrifugal force (N).

The value of centrifugal force and the time of the process must be provided in order to obtain a given degree of clarity.

In the clarification step of microbial suspensions, the tubular and disc centrifugal methods are often used. The tubular bowl centrifuge consists of a vertical cylinder, which rotates between 15 000 and 50 000 rpm, inside a stationary casting. The disc bowl centrifuge contains inverted cones or discs, which operate at 2000–7000 rpm. Due to the continuous and discontinuous operational behaviour of such centrifuges, the tubular centrifuge is applied to suspensions with a maximum of 30 g L^{-1} of cells, while in the disc, suspensions up to 250 g L^{-1} can be processed.[60]

A classic example of a clarification process used in Brazil is that of ethanol production by yeast *Saccharomyces cerevisiae*.[63,64] The viability of ethanol production is based on the efficiency of the clarifying operation, which depends on the recycling of yeast to the fermentation reactor and, consequently, the maintenance of high cell concentrations in the culture medium.[64]

Another relevant aspect for the appropriate process performance is the selectivity afforded by centrifugation, keeping bacteria in suspension while yeasts and other larger solids can sediment. The separation occurs due to the density difference between bacteria and yeast, the latter being removed from the supernatant due to their lower density. The partial removal of bacteria, the main contaminant, is a fundamental factor for an effective fermentation.[65]

2.2.3   Liquid/Liquid Extraction

The integration of fermentation bioprocesses with the initial step for the separation of the product can improve the product yield and facilitate the downstream processing. Liquid/liquid extraction seems to be the most promising among the approaches that have been used for this purpose. The extraction of metabolites produced *via* fermentation by liquid/liquid extraction has been the topic of many research activities and patents.[66] Their application in a large scale began in the middle of the 20th century. Since then, substantial advances in the industry have been observed, encompassing the most diverse sectors, from the extraction of noble metals, polymers and organic acids, among others.[67]

Since many biomolecules have narrow tolerance limits of pH, temperature, osmotic pressure, surface charges, among others, liquid/liquid extraction may be appropriate for these situations. For extracting proteins, the use of organic solvents is not appropriate since these molecules are susceptible to denaturation. Alternatively, proteins can be purified by using systems composed of two immiscible aqueous phases.[68]

In aqueous two-phase systems, the target molecule and impurities can be separated due to their differences in solubility between the liquid phases. Factors such as the surface properties of proteins, electrical charge, hydrophobicity and molecular mass must be taken into consideration.[61]

These systems may be formed by the combination of certain polymers, polyelectrolytes or also polymers in combination with low molecular weight solute. The most common systems are polyethylene glycol (PEG)/dextran, polypropylene glycol (PPG)/dextran, salt/PEG, and salt/PPG.[69]

For large-scale processing, systems consisting of PEG/dextran and PEG/salt are the most used since they are available commercially in large quantities, are non-toxic and capable of sterilization, having a greater range of application and adequate physical properties, particularly with respect to differences in density and viscosity. The addition of PEG and dextran to food is allowed in many countries, being allowed by the appropriate food and pharmaceutical legislation. The same can be said for certain types of citrates salts, phosphates and sulfates.[70]

2.2.3.1 Principles of Liquid/Liquid Extraction

Liquid/liquid extraction is a process that involves mass transfer between two immiscible or partially miscible liquids. The separation from a homogeneous liquid solution occurs by adding a liquid component, insoluble or partially soluble in the solution, the solvent, in which the component to be extracted from the solution, the solute, is preferably soluble. The solute diffuses in the solvent with a characteristic velocity until it reaches the equilibrium concentrations in each of the phases formed. This separation process is based on the distribution of the solute between the two phases and partial miscibility of the liquids.[71]

Knowledge of the phase equilibrium relationships is fundamental for quantitative analysis of extraction processes, which is based on thermodynamic principles and may be represented by diagrams established for law distribution of the constituents between the phases. In the liquid/liquid extraction, the effect of small changes of pressure on equilibrium is insignificant, restricting attention to the influence of temperature and concentration.[70]

Concentration of the components above critical concentration should be present in the solution to obtain two aqueous phases immiscible or partially miscible, which can be obtained from a phase diagram of the system of interest.

In a phase diagram, the vertical axis represents the mass composition of the molecule that has a higher concentration in the top phase (lower density phase) and the horizontal axis represents the composition of the molecule, presenting greater concentration in the bottom phase (phase of higher density). Compositions represented by points above the equilibrium curve contribute to the formation of two phases, and below the curve, to a single phase. To exemplify, the phase diagram for the PEG/potassium phosphate system at 20 °C and pH 7.0 is represented in Figure 2.7.[72]

Considering the system in equilibrium M, the composition of the top phase is represented by point T. The equilibrium concentrations of PEG and salt (on the bottom phase of the system) are given by the point B. The line connecting the point T to point B, through the point, M is called the tie tine.

If systems present an initial composition equal to that of the tie line, the same final composition may be found (top and bottom phase), but the volume ratio between the phases is different for each initial composition. The ratio between the volumes of the phases can be obtained through the ratio between the line segments MB and MT. The various tie lines are parallel to each other.[73]

The higher the molecular weight of the polymer, the lower the concentration needed for the formation of the two phases. Having this, the binodal curve is shifted to the direction of a single phase region.[74]

The type of salt, pH and temperature also influence this behaviour/curve. A pH reduction shifts the equilibrium curve significantly to the right. In systems composed of a polymer and salt, higher concentrations are necessary for the formation of two liquid phases.[75] In aqueous two-phase systems composed of two polymers, an increase in temperature promotes a higher miscibility between the polymers from system and, consequently, the curve

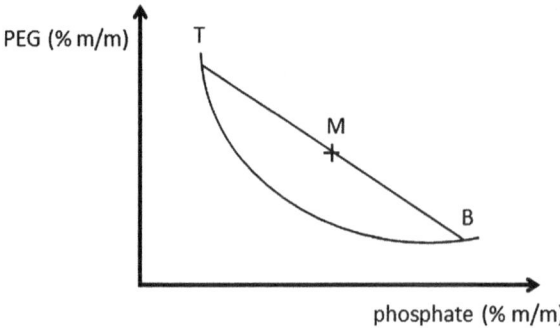

Figure 2.7   Schematic representation of the phase diagram of the PEG/phosphate system (% m/m).[60]

shifts and departs from the origin of the phase diagram.[75] In aqueous two-phase systems composed of a polymer and salt, an opposite effect is observed. With the increase in temperature, the concentrations of polymer and salt needed for the formation of two phases is reduced.

The extraction process in systems having two immiscible aqueous phases provides advantages such as the possibility of continuous operation at a large scale at room temperature; maintenance of the proteins in solution in the midst of polymers or salts, which are able to protect them from denaturation; and the possibility of elimination of the purification process steps for intracellular molecules.[60,65]

2.2.4 Cell Disruption

If biotechnological products of microbial origin are retained within the microbial cell after their production, an additional step to disrupt these cells may be required.[76] In spite of the existence of soluble microbial products (SMP), which are soluble organic compounds and easily released during the normal biomass metabolism,[77] *Escherichia coli* and *Saccharomyces cerevisiae*, the most common hosts used in large-scale manufacture of biotechnological products, do not excrete products to the medium.[78] In these situations, effective techniques for cell disruption are then required,[79] taking into account that microorganisms are more robust than is generally believed. The cell contents are separated in specific reservoirs from the outside environment by membranes, which provide mechanical strength to the cell and preserve its integrity. While animal cells are formed by fragile membranes and are easy to disrupt, yeast, bacterial and other fungal cells have rigid membranes, which need high shear stress to be disrupted.[76] Cell disruption should be carried out in order to (i) solubilize the maximum amount of product present in the cell while still maintaining maximum biological activity; (ii) to avoid possible secondary alteration of the product which will make it useless (denaturation, proteolysis and oxidation); and (iii) to limit the detrimental effects of the disruption stage on the following separation steps. To achieve all these objectives, cell disruption needs to be at the same time fast and efficient, since an incomplete disruption eventually results in low yield. Besides, the cell disruption technology used should be powerful enough to disrupt cell membranes or walls, and yet be gentle enough so that the materials inside the cell are not physically or chemically damaged.

The methods used to achieve cell disruption and to recover the aimed product can be classified as mechanical (high pressure homogenization, grinding balls, sonication), non-mechanical physical (osmotic shock, freezing and thawing, drying); chemical (alkalis, solvents, surfactants, supercritical CO_2 and acids) and enzymatic (enzymatic lyses or inhibition of cell wall synthesis),[80] as shown in Figure 2.8. These methods may be applied both individually and in combination. However, only some of these methods are able to achieve cell disruption at industrial levels (to be discussed in the following paragraphs). Overall, mechanical methods are often favoured for

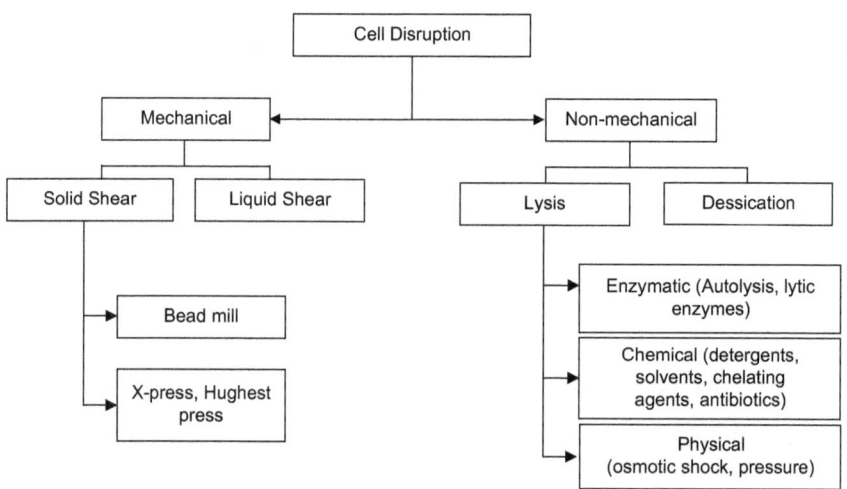

Figure 2.8   Cell disruption techniques. Adapted from Chisti and Moo-Young
(1986).[84]

large-scale cell disruption because they are easy to scale up and relatively
cheaper to operate.[81]

   Homogenization, a liquid shear method, was initially developed and used
in the dairy industry to prevent creaming during storage of milk. High-
pressure homogenization equipment operates by raising the pressure of a cell
suspension to approximately 1000 bar, and then releases this through a
specially designed valve assembly. The cells experience a range of forces, and
the disruption occurs *via* their interaction with the fluid and the solid walls of
the valve assembly.[82] Operation at the maximum attainable hydrodynamic
forces in the homogenizer will produce the maximum number of broken cells
and, consequently, maximize the release of the desired product.[83] On the
other hand, negative effects from this process can also be generated. Thermal
degradation of the product due to the excessive heating in the homogenizer
and the formation of cellular debris particles, which are small enough to
interfere with the DSP, are some examples of undesirable situations that can
be observed. When these products are intracellular macromolecules such as
proteins, antibodies, and DNA plasmids, there is also a potential risk of the
destruction of these molecules after they are released due to the same
hydrodynamic forces responsible for the disruption of the cell walls.[83]

   Ultrasonication is another liquid-shear method of cell disruption. Ultrasonic vibrations having frequencies greater than 18 kHz are able to disrupt
microbial cells in suspension.[84] The ultrasonic vibration could be emitted
continuously or in the form of short pulses. A frequency of 25 kHz is commonly used for cell disruption.[84] The duration of this procedure depends on
the cell type, the sample size and the cell concentration. The transmission of
sonic waves creates a continuous cycle of microbubble cavitation in the
suspending medium. These cavities or small bubbles of dissolved gases or

vapours arise from the alteration of pressure in liquids. The microbubbles form during the rarefaction phase of the sound wave and collapse during the compression phase. The microbubble implosion releases a large amount of mechanical energy as shock waves. The continuous bombardment of cells by these violent waves eventually shear the cell membrane/wall and disrupts the cells.[84]

One of the most efficient techniques for physical cell disruption is the grinding of the cells in a ball mill (solid shear). These mills consist of either a vertical or a horizontal cylindrical chamber with a motor-driven central shaft supporting a collection of off-centred discs or other agitating elements.[85] In this process, cells are agitated in suspension with small abrasive particles. Cells break because of shear forces, grinding between beads and collisions with beads. The beads disrupt the cells to release biomolecules but do not break the target molecules. The kinetics of biomolecule release by this method is a first-order process.

Chemically, cell disruption can be achieved through the use of detergents, solvents, alkalis, chelating agents and supercritical CO_2. Organic solvents such as acetone, chloroform or dichloromethane mainly act on the cell membrane by solubilizing its phospholipids and by denaturing its proteins. Some solvents like toluene are known to disrupt fungal cell walls; however, it is highly toxic and can not be used when the product is intended for pharmaceutical use. The limitations of using organic solvents are similar to those with detergents, *i.e.* they require the removal of these solvents and may cause denaturation of proteins. However, organic solvents are easier to remove than detergents, on account of their volatility.

Solvent extraction process is a well-established chemical method to recover polyhydroxyalkanoates (PHAs) from the cell biomass. This method is also used routinely in the laboratory because of its simplicity and rapidity. Firstly, the cell membrane permeability is modified, allowing release and solubilization of PHA.[86] Then, a non-solvent precipitation is performed.[87] Extraction of PHA with solvents such as chlorinated hydrocarbons, *i.e.* chloroform, 1,2-dichloroethane or some cyclic carbonates like ethylene carbonate and 1,2-propylene carbonate is common.[88] Short-chain ketones such as acetone is the most prominent solvent, especially for the extraction of medium-chain-length polyhydroxyalkanoates.[86,87] Although extraction with solvent seems promising for PHA extraction, disadvantages need to be considered. The target material can be degraded[88] and, in the extraction of proteins, they can be denatured by the solvent and then lose their activity.

In view of the above-mentioned disadvantages for systems using solvents and that the high pressure homogenizer may provide the release of the desired product together with the release of other contaminating molecules and cellular debris, enzymatic methods have been proposed recently in order to overcome these disadvantages.[89] Monks *et al.* have tested different treatment protocols (enzymatic, chemical and ultrasound-assisted) to disrupt the cell of *Sporidiobolus salmonicolor* (CBS 2636) and low amount of carotenoids have been recovered when the enzymatic treatment is not used – approximately 26%.[90] Ultrasound-assisted cell disruption alone or combined had lower performance. Besides the high performance, enzymatic treatments uses green chemistry principles to guide new products/process development, which makes it advantageous. Enzyme systems can provide biological specificity to the process of cell disruption and a controlled lysis and release of product from microbial cells.[91] These methods may be used alone for releasing intracellular proteins, polysaccharides, particulate inclusions and for wall and membrane associated materials[89,91,92] or they can be associated with other methods aiming to increase the selectivity, the rate and yield of extraction, and also to minimize product damage. The industrial use of enzymes for releasing specific cell proteins started with the development of expression systems for large recombinant proteins that cannot be secreted by the cells.[92]

Since different methods can be used to disrupt cells, some considerations should be taken when selecting them, for instance, some disruption methods which work very well for animal tissue may not work at all for microbial cells. In this situation, the type of cells and their growth and storage history should be carefully analyzed.[93] Ultrasound methods generate heat and thus can destroy organelles, affecting the configuration of biological molecules. Shearing systems like blenders and ball mills can shatter the cell contents as well as the cell walls. Furthermore, the amount of energy required to disrupt cells depends on the type of organisms. Important considerations also include the volume and number of samples, the disruption time, the scale-up potential, the availability of equipment and the economic feasibility.

2.2.5   Polishing

In the polishing stage, the objective is to achieve high purity by removing any remaining trace impurities or closely related substances. It is a critical step if clinical grade material is envisaged.[94] Since high resolution and recovery are expected in this step, chromatographic methods have been the most used.

2.2.5.1   Chromatographic Approaches

The major limitations in the downstream processing are found in the selective purification steps, currently dominated by chromatography, which accounts for more than 70% of the downstream costs.[95] Chromatography is still the major tool on all levels of the downstream processing, from the first capture to the final polishing step due to its advantages such as high separation efficiency in just one step (hundreds or even thousands of theoretical plates while extraction or filtration membranes are limited to a few stages), allowing the resolution of complex mixtures with very similar molecular properties; packing columns presenting high capacity adsorbents, which are ideal for capturing molecules from dilute solutions; stationary phase can be easily regenerated.[96] The difficulty of scale-up is its principal disadvantage.

Moreover, there is no economy of scale with robust and reliable columns because the additional costs of resins, buffers and other consumables outstrips any savings made by increasing the productivity.[97]

Conventionally, chromatographic separations of biotechnological products are based on a single physical property, such as charge (ion-exchange chromatography), hydrophobicity (hydrophobic interaction chromatography), size (size exclusion chromatography), specific interactions (affinity chromatography), or metal-chelating groups (immobilized metal ion affinity chromatography).[98] The selection of chromatographic method will depend on downstream process step and also physicochemical characteristics of the material to be purified. Overall, adsorptive chromatography is used in intermediate purification/concentration while ion-exchange chromatography and size exclusion chromatography is used in polishing.[94]

Ion-exchange chromatography (IEX) is one of the most widely used chromatographic steps in downstream purification of biotechnological products. Protein retention in ion-exchange chromatography is a complex function of stationary and mobile phase effects.[99] Although electrostatic interaction forces represent the primary mode of retention in ion-exchange systems, non-specific interactions such as van der Waals and hydrophobic interactions can also play an important role in determining the selectivity in these systems.[100] Anion-exchange chromatography has been widely used during the purification of monoclonal antibodies to remove high-molecular-weight contaminants such as DNA and viruses. Such molecules do not readily diffuse into the pores of traditional resins, resulting in mass transfer resistance and lower efficiency. Absorbers membrane has recently been used to replace the traditional columns, which would provide hydrodynamic benefits. They are able to operate at much greater flow rates than columns, reducing buffer consumption and shortening the overall process time.[97]

Hydrophobic interaction chromatography (HIC) is based on interactions between hydrophobic parts of biomolecules and hydrophobic ligands of the adsorbent, which are highly dependent on the salt concentration of the buffer used in the mobile phase. There are no permanent bonds in buffers with low salt concentrations, because hydrate layers around the hydrophobic parts of selective ligands like phenyl- or butyl- and corresponding biomolecules surface parts avoid bonds. On the other hand, hydrophobic parts can establish bonds when salt concentration is increased because the water molecules leave the layers around the hydrophobic parts.[101] Membrane adsorbers and monolith supports usually are used as stationary phase, and their micro- or macroporous surfaces can be derivatized with functional ligands able to improve the separation efficiency by increasing solute selectivity. Furthermore, membrane and monolith supports also bear minimum diffusive mass transfer to the surface, which combined with low pressure drops, even at higher flow rates, improves productivity.[102] Matrices presenting larger porous size have been recently proposed to purify virus-like particles regarding their larger size compared to proteins, which provide lower diffusional mass transfer resistances.[94] In the purification process of recombinant proteins, this technique has as an advantage in the fact that the chromatographic conditions are very close to the physiological conditions of the human body (neutral pH, an aqueous salt solution, and room temperature), all of which are favourable to the maintenance of the proteins' bioactivity. On the other hand, classical commercial hydrophobic adsorbents have been inadequate for downstream processing of recombinant proteins because of their high hydrophobicity and thus different modifications in the stationary phase have been proposed.[103] This technique has also been successfully used for the separation of plasmid DNA from endotoxins and single-stranded nucleic acids. In this situation, reductions in the salt concentration of the mobile phase are performed to reduce the hydrophobic interactions and to improve the separation efficiency.[104]

Affinity chromatography (AC) is a separation method based on the formation of a specific and reversible complex between the target and the ligand. The ligand is often used in an immobilized form (insoluble matrix). These steps are usually employed as the capture chromatographic steps in the downstream process due to their high selectivity and ability to concentrate and separate the product readily. This also allows the clearance of any leached affinity ligand through subsequent polishing steps in the process.[96] Immunoaffinity chromatography is a principal affinity-based separation technique, characterized by the interaction between an antigen and a matrix-bound antibody that is able to recognize a structural component of a studied material (for example, it is able to recognize a surface component from a viral particle). This procedure can result in high yield in a single purification step, which simplifies the following downstream process. On the other hand, the high costs associated with antibody purification and immobilization may be prohibitive for industrial-scale use.[105,106] It has been the method of choice for the large-scale purification of plasmid DNA in view of the complexity of biomolecules present in plasmid DNA-containing extracts and their structural and chemical similarities with those from impurities.[107]

In size-exclusion chromatography (SEC), molecular size, or more precisely, molecular hydrodynamic volume, governs the separation process. That is, as a mixture of solutes of different size pass through a column packed with porous particles, the molecules that are too large to penetrate the pores of the packing elute first. On the other hand, smaller molecules able to penetrate or diffuse into the pores, elute at a later time or elution volume.[108] A variation of this technique is a size-exclusion reaction chromatography (SERC), a method for controlling reactions that combines reaction and separation in a single unit operation so as to obtain a high yield of a product with a specific molecular weight. It utilizes a moving reaction zone within a size-exclusion chromatography column to control the time of contact between reactants, to selectively remove products from the reaction zone, and to selectively inhibit reactions based on molecular size. The applicability of SERC is not restricted to protein PEGylation reactions but could potentially be used for any reaction that results in a change in molecular size, such as those involving glycosylation, polymerization, or cleavage reactions, fusion proteins, poly- or oligo-peptides, amino acids, carbohydrates, inclusion bodies, microbial cells and viral particles.[109]

Immobilized metal ion-affinity chromatography (IMAC) was first established as a technique to fractionate proteins on solid supports based on their differential affinity towards immobilized metal ions. This differential affinity derives from the coordination bonds formed between metal ions and certain amino acid side chains exposed on the surface of the protein molecules. Since the interaction between the immobilized metal ions and the side chains of amino acids has a readily reversible character, it can be used for adsorption and then be disrupted under mild conditions, usually by adding a competing agent.[110] The use of this technique to purify recombinant proteins containing a short affinity-tag consisting of poly-histidine residues represent its main application (the most widespread and versatile strategy used to purify recombinant proteins).[111]

Due to the structural complexity of some biomolecules and the purity criteria required by regulatory agencies, these single methods have not been able to efficiently provide the separation for many biotechnological products. Thus, efforts using multimodal chromatography have been made in the last decades, where two or more physicochemical properties are used to enhance the specificity of the interactions between the protein and the ligand on the chromatographic matrix.[98] For example, the platform approach for the purification of monoclonal antibodies usually includes three chromatographic steps: a protein A chromatography capture step and two chromatographic polishing steps.[95]

Predictive quantitative structure – property relationship (QSPR) models involving regression algorithms have also been proposed in the last decade for the prediction of chromatographic behaviour of proteins.[112] The practical application of this technique (QSPR) in a typical downstream bioprocessing setup would involve the generation of models to predict the chromatographic behaviour of the product of interest and the key impurities in a given biological mixture. Moreover, computational experiments may also be carried out by varying operational parameters in that QSPR models are able to predict the resolution of the resultant separation. Thus, this strategy can enable the in silica design and optimization of chromatographic separations of bioprocess mixtures.[100]

2.2.6 Non-chromatographic Approaches for Purification

Chromatographic separations have presented an unquestionable potential to obtain high-purity compounds; however, the increasing productivity of biotechnological products to supply the demand together with the high cost of chromatography processes are pushing the development of more-efficient and cost-effective separation and purification methods.[95] Although chromatographic methods such as column chromatography or high-pressure liquid chromatography are highly specific, they can handle only small amounts of feed at a time.[68,113] On the other hand, a variety of biotechnological-derived products have been successfully purified without the use of chromatography, such as plasmid DNA,[114,115] viruses[116] and polysaccharides.[117]

A good example of a bulk separation are the aqueous two-phase systems (ATPS). They can be used as an alternative and efficient approach for purification of biomolecules by partitioning between two liquid phases. ATPS result from the incompatibility of polymers, either between two polymers in water or a polymer solution with a salt solution.[68] In this system, the cells are considered to be immobilized on to one of the phases of the ATPS and the required product is made to partition into the other phase by proper manipulation of the system.[68] ATPS can be used to purify proteins,[118,119] cells,[120] viruses particles[121] and plasmid DNA.[120]

Another bulk separation is three-phase partitioning (TPP), a protein separation method based on heterogeneous systems consisting of *t*-butanol, ammonium sulfate and water. *t*-Butanol binds to the precipitated proteins, thereby increasing their buoyancy and causing the precipitates to float above the denser aqueous salt layer. Optimum pH, temperature, ammonium sulfate and *t*-butanol concentrations can selectively precipitate proteins at the interface between the organic and aqueous phase. Kosmotropy, salting out, co-solvent precipitation, isoionic precipitation, osmolytic electrostatic forces, conformation tightening and protein hydration shifts contribute to this protein precipitation.[122–126] TPP is a simple, inexpensive, scalable, and rapid procedure, works at room temperature, and the chemicals used in the process can be recycled. It does not use polymers which have to be removed later. *t*-Butanol is normally completely miscible with water (b.p. 84 °C) and is much less flammable than hexane, methanol or ethanol which are used in conventional extraction.[127]

Precipitation and crystallization are unit operations that generate a solid from a supersaturated solution. They can be distinguished based on the speed of the process and the size of the solid particles produced.[128] Precipitation has been used to purify proteins and nucleic acids while crystallization has been used for the purification of products ranging from industrial enzymes (*e.g.* glucose isomerase) to drugs (*e.g.* insulin). It is a powerful technique for low-cost purification since it uses low temperatures and energy consumption, with high purification yield in one single step. The challenge remains to develop both the theory and screening techniques to be able to establish a robust zone of high yield crystallization without excessive empirical experimentation.[128]

Among the methods of field-based separations, membrane filtration is one of the most representative, which includes microfiltration, sterile filtration and ultrafiltration.[58,129–131] Microfiltration is used for clarification and sterile filtration, and ultrafiltration for protein concentration and buffer exchange.[58]

Adsorptive separations can be exemplified by the monolith system. Macroporous monoliths are stationary phases that can be prepared in a variety of shapes and dimensions using relatively straightforward polymerization chemistry and which can be derivatized with traditional chromatography ligands.[128,132,133]

2.3   Examples of Downstream Processing of Different Products

Each product obtained through biotechnology processes requires specific steps for its recovery. Below there are some examples of important materials produced through white biotechnology.

2.3.1   Recovery of Biopolymers or Bio-based Monomers for Further Polymerization

2.3.1.1   *Polyhydroxyalkanoates (PHAs)*

PHAs are a family of thermal biopolyesters with over 100 different monomer structures, which can be synthesized by different bacteria. Their different properties arise chemically, either from the length of the pendant groups that extend from the polymer backbones, or from the distance between the ester linkages in the polymer backbones.[134] PHAs can be used to produce several medical devices, such as tissue coating sutures, fully absorbable stents, bone, tubing and other prostheses, tissue regenerating devices, wound dressings, absorbable medical devices with higher strength and flexibility, and drug delivery.[135]

PHA is accumulated intracellularly in Gram-negative bacterial strains and normally its recovery after the fermentation include several steps – briefly these are: (i) the separation of cells from the fermentation broth by centrifugation; (ii) after that, the bacterial cells are pre-treated by heat, freeze dried, or salted, before extraction to avoid polymer degradation; (iii) the PHA is therefore extracted, normally by using chlorinated solvents or other methods such as enzymatic digestion or mechanical cell disruption; and (iv) PHA purification. The process of PHA recovery is shown in Figure 2.9.

The use of solvents to recover PHA is one of the oldest and commonly used methods. Firstly, it affects the cell membrane permeability and then it dissolves the PHA.[136] However, PHAs may be contaminated by lipopolysaccharides (LPS) during the extraction process if solvents are used. LPS is an component from the cell wall of the Gram-negative bacteria, which is pyrogenic in nature.[137] For medical purposes, PHAs need to be highly pure, especially with respect to pyrogenic compounds. LPS are considered the main source of pyrogenic contamination.[138]

In such situations, downstream processing parameters such as biomass pre-treatment, the kind of extraction method employed, and polymer purification steps become very crucial in removing these contaminating lipopolysaccharides.[137] However, some methods used to remove LPS from polymers, including the treatment with ozone, hydrogen peroxide, sodium hypochlorite and NaOH can reduce the molecular weight of the extracted PHAs. For example, when sodium hypochlorite digestion was carried out, severe degradation of the polymer, with up to 50% reduction in molecular weight, occurred. Solvent extraction of medium-chain-length PHAs have

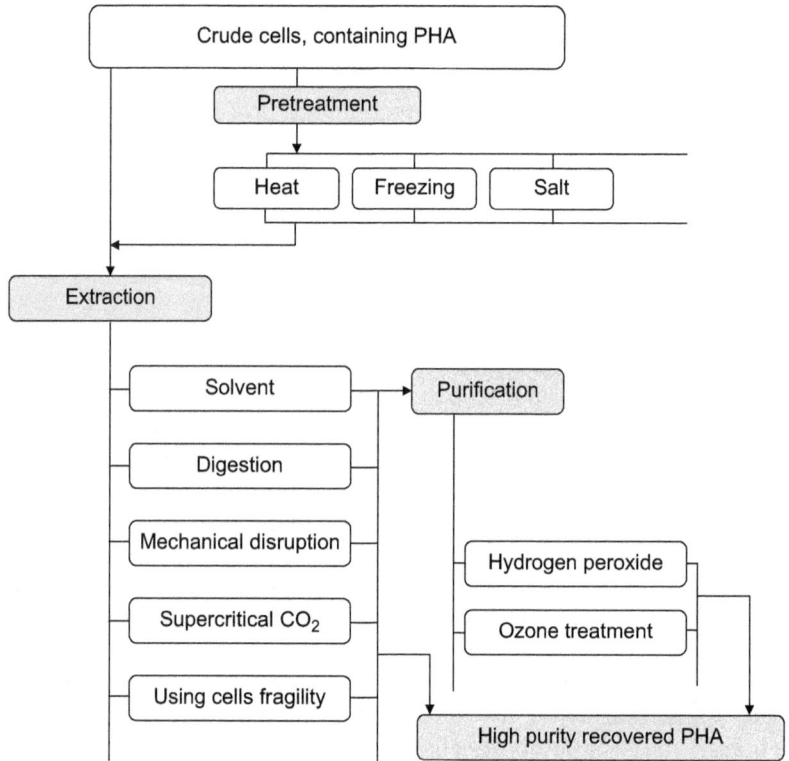

Figure 2.9   Purification strategy for the polyhydroxyalkanoates. The recovery of
polyhydroxyalkanoates is composed of three steps: pre-treatment,
extraction and purification.
Adapted from Jacquel *et al.* (2008).[136]

been carried out using a wide range of solvents, because, unlike short-chain
length PHAs, they are soluble in a wider range of solvents.[137]

Furrer *et al.* have tested different solvents to recover PHA and tested the
influence on the toxicity of the polymer. The results showed significant
differences in endotoxicity of PHA due to the use of different solvents. Ethyl
acetate, acetone, tetrahydrofuran (THF), 2-propanol and *t*-butyl methyl ether
(MTBE) extracts showed a much lower endotoxin content than the extracts
with methylene chloride (CH_2Cl_2) and hexane.[138]

The use of oxidizing agents such as hydrogen peroxide or benzoyl peroxide
has also reduced the endotoxin content to less than 20 endotoxin units/gram
of PHA, and this PHA did not result in an acute inflammatory response when
implanted in animal models.[139]

2.3.1.2   *Lactic Acid and Polylactic Acid (PLA)*

Lactic acid (2-hydroxypropionic acid), CH_3–$CHOHCOOH$, is a simple chiral
molecule which exists as two enantiomers, L- and D-lactic acid, differing in

their effect on polarized light. The optically inactive meso form is an equimolar (racemic) mixture of D-(−) and L-(+) isomers.[140–142] It ranks among high-volume chemicals produced with microorganisms, with an annual world production volume of about 370 000 MT.[143]

Lactic acid has been produced in industrial scales since the end of the nineteenth century and is mainly used in the food industry, cosmetics, pharmaceuticals and animal feed, particularly due its action as an acidity regulator. Moreover, it is the monomeric precursor of PLA,[144] which can be obtained either by carbohydrate fermentation or by common chemical synthesis.[145]

The purification step is the most important in the commercial production of lactic acid, which needs to meet the quality standards requirements for specific applications. The conventional fermentation process produces lactate salt, as result of pH neutralization, that must be precipitated and reacidified by a mineral acid such as sulfuric acid.[146]

For the production of polymers or their derivatives, the technology for producing the dilactide (the internal diester) is critically important.[147] The processes include a multistage evaporation followed by polymerization to a low molecular weight prepolymer, which is then catalytically converted to the dilactide. The dilactide is purified in a distillation system by partial condensation and recycling. This diester can be used to synthesize high molecular weight polymers and copolymers.[147]

Fermentation-derived lactic acid can be separated by several recovery processes, which include calcium precipitation, solvent extraction and electrodialysis.[148] Other recovery techniques have also been reported such as direct distillation, adsorption, liquid surfactant membrane extraction, chromatographic approaches, ultrafiltration, reverse osmosis, drying, conventional electrodialysis as well as bipolar membrane electrodialysis.[146]

2.3.2 Pharmaceutical Applications: Viral Vectors, Plasmid and Vaccine Manufacturing Processes

Although the focus of this chapter is limited to downstream processing, it is worthwhile keeping in mind that process development for production of viral gene therapy vectors, vaccines and other biopharmaceutical products, in fact, consists of three integrated stages: upstream processing (selection of appropriate producer cell culture lines), defining optimal growth conditions, and downstream processing (isolation and purification steps).[106,149] It is desirable that upstream systems are designed in parallel with downstream technologies regarding their interdependence (downstream processing is greatly affected by the impurities and contaminants present in the process streams) and that as much as 70% of production costs may be associated with the downstream processing and purification operations.[149] Well-defined upstream processes could reduce the number of steps required to purify these materials and consequently reduce the production costs.

Critical issues such as product degradation or diminished stability should be overcome as early as possible in the process and purification methods should be selected regarding limitations of the molecule/product.

The downstream processes should be designed to accommodate the best compromise between cost, throughput and purity needs, so meeting the desired purity and potency.[94] In this sense, chemical, biochemical and physics assays should be used to ensure that the processes may be validated so that the biopharmaceutical products present the intended features in terms of safety, potency and purity.[150] For example, controls on the percentage of dimers, oligomers and higher aggregate forms are often required during the recombinant protein processing. The protein aggregation can induce immune reactions, cause other side effects or it may be related to stability problems: *e.g.* constitute seeds for precipitation and reducing the shelf life of a product.[96] Vaccine products can be inadvertently denatured or rendered irreversibly inactive due to mechanical damage or by deleterious chemical effects during processing. Hence, the selection of equipment and solvents is important in minimizing product loss, and for maintaining an efficient and optimized manufacturing process.[151]

As previously mentioned, the reduction in the number of steps during production is relevant to reduce the production costs. During virus-like particle production, for example, the downstream processing depends on whether particles are released to the extracellular medium. If these particles are not efficiently secreted, a cell lysis or other extraction step might be required before the actual clarification step (a step designed to efficiently remove cell debris and large aggregates). Thus, biotechnology industries have made substantial efforts to design a clone compatible with an efficient secretory pathway.[94]

Purity requirements depend on the intended use of the biopharmaceutical, dose and risk-benefit ratio. The selection of purification method should consider physicochemical characteristics of the particle to be isolated. For viral particles it is relevant to consider the influence of isoelectric point (pI), surface hydrophobicity, the presence or absence of an envelope, hydrodynamic diameter and lability of virus particles. Viruses should not have their level of infectivity reduced after a purification stage when they are used as vectors, and tests should be carried out to check this.[150] During the process of vaccine manufacturing, the production of virus-like particles is dependent on physicochemical parameters such as pH, ionic strength and medium temperature. Type and concentration of chelating agents also seem to affect the protein macrostructure stability and thus should be observed.[152]

A purification process for biopharmaceuticals must ensure that contaminants from different sources such as the bioreactor as well as from the cell are removed from the final product to obtain the desired product quality and satisfy the regulatory requirements.[152] A variety of purification methods have been proposed, such as density-gradient ultracentrifugation, ultrafiltration, precipitation, two-phase extraction systems and size exclusion chromatography (Table 2.1), each one being used in specific stages.

Table 2.1 Advantages, disadvantages and applications of most often used methods in downstream processing of biopharmaceutical products.

Method	Advantages	Disadvantages	Characteristics from material to be purified	Ref.
Ultracentrifugation	It is a versatile tool for most of the biotechnology molecules since it is included in industrial plant of various biotechnology industries. One of the most commonly used techniques in the concentration step.	Non-scalable, tedious and very labor intensive. Low recovery yield and often several impurities are still present at the end of the process. Toxic and mutagenic reagents are used in some situations. Virus particle may suffer a loss of infectivity depend on required time period for the separation. Initial investment is four times higher than the ultra-filtration method.	Rate-zonal centrifugation may be used for preparation of larger amounts of DNA nanostructures. Inappropriate for viral vectors.	105, 150–152
Flocculation/ precipitation	No complex equipments, low-cost production.	It should be combined with a centrifugation method. Impurities or polymers may co-precipitate along with particles to be purified. Handling of corrosive chemicals, heavy wastage, toxicity and biomass yield is very reduced. Optimum operating conditions (type and concentration of the precipitant agent, temperature, aging time, agitation, pH) for the precipitation are required in various situations.	Highly cost-effective for purification of some enzymes.	105, 150–151
				105, 150

Table 2.1 (*Continued*)

Method	Advantages	Disadvantages	Characteristics from material to be purified	Ref.
Membrane-based tangential flow filtration (ultrafiltration and microfiltration)	No wastage, toxicity and biomass yield is high (compared to centrifugation) and cassette can be reused. Initial investment is less than centrifugation (2–3 times cheaper).	Membrane fouling and loss of viral infectivity due to high pressures may be observed in specific situations. Given that the separation is based on size differences, large molecular weight transduction inhibitors may be concentrated with the viral particles (reducing viral transduction efficiencies).	Especially applicable for sensitive molecules. It can be used in purification of virus for vaccine production given that the integrity of virus structural is not compromised. Pore size of the membrane may be adjusted according to the purpose of separation.	
Size exclusion chromatography	It is mainly applied as a polishing step.	Difficulties in up-scaling and a low throughput. Separation from high molecular weight components such as proteoglycans or cellular genomic DNA is difficult.	Virus and crude cell lysates can be separated. Alternatively, it can be used when particles to be purified and impurities present the same electrostatic properties. It has capability to refold a variety of proteins.	105, 149–152

The number of steps is related to the complexity of the molecule/product to be isolated and its intended purity level. In plasmid-DNA preparations, the clarification and concentration is performed after the lysis step to remove cell debris and structurally unrelated impurities (proteins and low molecular weight nucleic acids), while simultaneously concentrating and conditioning the plasmid-DNA preparations for the next step. In this last step, chromatography is usually used to separate supercoiled plasmid DNA from structurally related impurities such as relaxed and denatured plasmid DNA, genomic DNA (gDNA), high molecular weight RNA and endotoxins.[150]

Nowadays, significant advances in the design and construction of safer and more efficient processes are taking place. The identification of more specific ligands in affinity-based purification systems would increase the separation efficiency and thus reduce costs (fewer stages would be necessary for purification). Strategic considerations regarding the design of appropriate separation systems in tandem are also suggested.[106]

2.4 Conclusions

An efficient and cost-effective biotechnological process depends largely on the downstream processing that ensures the purity and quality of the desired products. The majority of biotechnology processes for producing valuable products involve the purification of different molecules from a variety of sources such as bacteria, yeast and mammalian cell culture fluids, or extracts from naturally occurring tissue. Since many biomolecules have narrow tolerance limits of pH, temperature, osmotic pressure, surface charges, *etc.* the extraction and isolation techniques should be specific and compatible to the product. Downstream purification problems for biomanufacturers finally seem to be improving, with several studies devoted to the development of new methodologies or the improvement of original ones. Each additional step in the recovery process will have an economic impact by increasing operational cost and process time, and also by causing loss in product yield. Careful selection and combination of suitable unit operations during the design phase may reduce the number of steps needed. Understanding the importance of having a holistic approach to properly design an experiment from the beginning, including the downstream process, is now well understood. To provide the quantity and quality of product required using the least number of steps, in the most cost-effective manner, it is crucial to integrate all parts of the process.

References

1. M. Gavrilescu and Y. Chisti, *Biotechnol. Adv.*, 2005, **23**, 471–499.
2. D. R. Headon and G. Walsh, *Biotechnol. Adv.*, 1994, **12**, 635–646.
3. *Focus on Catalysts*, 2004, **2004**, 2–3.
4. C. Lu, J. A. Napier, T. E. Clemente and E. B. Cahoon, *Curr. Opin. Biotechnol.*, 2011, **22**, 252–259.

5. R. H. Wijffels, O. Kruse and K. J. Hellingwerf, *Curr. Opin. Biotechnol.*, 2013, **24**, 405–413.
6. R. Wohlgemuth, *New Biotechnol.*, 2012, **29**, 165.
7. U. Gottschalk, in *Disposable Bioreactors*, eds. R. Eibl and D. Eibl, Springer, Berlin Heidelberg, 2010, vol. 115, pp. 171–183.
8. S. Y. Lee, D. Mattanovich and A. Villaverde, *Microbial Cell Factories*, 2012, **11**, 156.
9. R. Hatti-Kaul, U. Törnvall, L. Gustafsson and P. Börjesson, *Trends Biotechnol.*, 2007, **25**, 119–124.
10. M. Kircher, *New Biotechnol.*, 2012, **29**, 243–247.
11. J. C. Philp, R. J. Ritchie and J. E. M. Allan, *Trends Biotechnol.*, 2013, **31**, 219–222.
12. P. L. Rogers, Y. J. Jeon and C. J. Svenson, *Process Safety and Environmental Protection*, 2005, **83**, 499–503.
13. M. Hasheminejad, M. Tabatabaei, Y. Mansourpanah, M. K. Far and A. Javani, *Bioresour. Technol.*, 2011, **102**, 461–468.
14. E. Jain and A. Kumar, *Biotechnol. Adv.*, 2008, **26**, 46–72.
15. M. Holzer, *Biopharm Int.*, 2011, **24**, 48–53.
16. M. R. Pursell, M. A. Mendes and D. C. Stuckey, *Biotechnol. Prog.*, 2009, **25**, 1686–1694.
17. S. M. Wheelwright, *J. Biotechnol.*, 1989, **11**, 89–102.
18. W. L. Tang and H. Zhao, *Biotechnol. J.*, 2009, **4**, 1725–1739.
19. H.-J. Huang, S. Ramaswamy, U. W. Tschirner and B. V. Ramarao, *Sep. Purif. Technol.*, 2008, **62**, 1–21.
20. L. Y. Jiang and J. M. Zhu, *Wiley Interdisciplinary Reviews: Energy and Environment*, 2013.
21. M. Kalyanpur, *Mol. Biotechnol.*, 2002, **22**, 87–98.
22. C. J. Dowd and B. Kelley, in *Comprehensive Biotechnology (Second Edition)*, ed. M.-Y. Murray, Academic Press, Burlington, 2011, pp. 799–810.
23. Y. Shan, Y. Zheng, F. Guan, J. Zhou, H. Zhao, B. Xia and X. Feng, *Acta Biochim. Biophys. Sinica*, 2013, **45**, 649–655.
24. J. Arnau, C. Lauritzen, G. E. Petersen and J. Pedersen, *Protein Expression Purif*, 2006, **48**, 1–13.
25. F. Campos, G. Guillén, J. L. Reyes and A. A. Covarrubias, *Protein Expression Purif.*, 2011, **80**, 47–51.
26. C.-T. Lin, P. A. Moore, D. L. Auberry, E. V. Landorf, T. Peppler, K. D. Victry, F. R. Collart and V. Kery, *Protein Expression Purif.*, 2006, **47**, 16–24.
27. R. Ghosh, *Principles of Bioseparations Engineering*, 2006, 1–12.
28. R. Hatti-Kaul, in *Industrial Biotechnology – Sustainable Growth and Economic Success*, eds. W. Soetaert and E. J. Vandamme, Wiley-VCH, Ghent, 2010, pp. 279–318.
29. N. K. Prasad, *Downstream Process Technology: A New Horizon in Biotechnology*, Asoke Ghosh, New Delhi, 2010.
30. B. Burghoff, *J. Biotechnol.*, 2012, **161**, 126–137.

31. J. Merz, B. Burghoff, H. Zorn and G. Schembecker, *Sep. Purif. Technol.*, 2011, **82**, 10–18.
32. P. J. Martin, H. M. Dutton, J. B. Winterburn, S. Baker and A. B. Russell, *Chem. Eng. Sci.*, 2010, **65**, 3825–3835.
33. C. Santana, L. Du and R. Tanner, in *Biotechnlogy Vol. IV*, 2013.
34. G. D. Najafpour, in *Biochemical Engineering and Biotechnology*, Elsevier, Amsterdam, 2007, pp. 390–415.
35. Q. Wen, Z. Chen, C. Wang and N. Ren, *J. Environ. Sci.*, 2012, **24**, 1744–1752.
36. A. L. van Wezel, G. van Steenis, P. van der Marel and A. D. M. E. Osterhaus, *Rev. Infectious Diseases*, 1984, **6**, S335–S340.
37. C. Marin-Muller, A. Rios, D. Anderson, E. Siwak and Q. Yao, *J. Virol. Methods*, 2013, **189**, 125–128.
38. A. Rios, J. Quesada, D. Anderson, A. Goldstein, T. Fossum, S. Colby-Germinario and M. A. Wainberg, *Virus Res.*, 2011, **155**, 189–194.
39. I. C. Kusters, J. Matthews and J. F. Saluzzo, in *Vaccines for Biodefense and Emerging and Neglected Diseases*, eds. D. T. B. Alan and R. S. Lawrence, Academic Press, London, 2009, pp. 147–156.
40. F. Stauffer, T. El-Bacha and A. T. D. Poian, *Recent Patents on Anti-Infective Drug Discovery*, 2006, **1**, 291–293.
41. A. S. Sattar and S. Springthorpe, in *Principles and Practice of Disinfection, Preservation, and Sterilization*, eds. A. D. Russell, W. B. Hugo and G. A. J. Ayliffe, Blackwell Science, Oxford, 1999.
42. J.-Y. Maillard, *Rev. Med. Microbiol.*, 2001, **12**, 63–74.
43. A. D. Russell, *J. Antimicrob. Chemother.*, 2003, **52**, 750–763.
44. World Health Organization, *Guidelines on Viral Inactivation and Removal Procedures Intended to Assure the Viral Safety of Human Blood Plasma Products*, 2004.
45. M. A. Collis, B. K. O'Neill, C. J. Thomas and A. P. Middelberg, *Bioseparation*, 1996, **6**, 55–63.
46. C. D. Lytle and J. L. Sagripanti, *J. Virology*, 2005, **79**, 14244–14252.
47. R. P. Sinha and D.-P. Hader, *Photochem. Photobiol Sci.*, 2002, **1**, 225–236.
48. K. Sirikanchana, J. L. Shisler and B. J. Marinas, *Water Res.*, 2008, **42**, 1467–1474.
49. A. Sharma, P. Gupta and R. K. Maheshwari, *Virology J.*, 2012, 9.
50. M. T. Aspelund and C. E. Glatz, *J. Membrane Sci.*, 2010, **365**, 123–129.
51. P. Richardson, J. Molloy, R. Ravenhall, I. Holwill, M. Hoare and P. Dunnill, *J. Biotechnol.*, 1996, **49**, 111–118.
52. Z. L. Boynton, J. J. Koon, E. M. Brennan, J. D. Clouart, D. M. Horowitz, T. U. Gerngross and G. W. Huisman, *Appl. Environ. Microbiol.*, 1999, **65**, 1524–1529.
53. H. F. Liu, J. F. Ma, C. Winter and R. Bayer, *Mabs*, 2010, **2**, 480–499.
54. G. M. Lee, A. Varma and B. O. Palsson, *Biotechnol. Bioeng.*, 1991, **38**, 821–830.
55. L. Reeves and K. Cornetta, *Gene Therapy*, 2000, **7**, 1993–1998.
56. G. Robic and E. A. Miranda, *Biotechnol. Bioprocess Eng.*, 2011, **16**, 777–784.

57. N. Singh, K. Pizzelli, J. K. Romero, J. Chrostowski, G. Evangelist, J. Hamzik, N. Soice and K. S. Cheng, *Biotechnol. Bioeng.*, 2013, **110**, 1964–1972.

58. R. van Reis and A. Zydney, *Curr. Opin. Biotechnol.*, 2001, **12**, 208–211.

59. R. Gomide, *Operações unitárias - Vol 2: Separações Unitárias*, São Paulo, 1980.

60. W. Schmidell, U. A. Lima, E. Aquarone and W. Borzani, *Engenharia Bioquímica*, Blucher, São Paulo, 2001.

61. C. J. Geankoplis, *Transport Processes and Unit Operations*, 4th edition, Prentice Hall, Upper Saddler River, 2003.

62. M. A. Cremasco, *Operações Unitárias em sistemas particulados e fluidomecânicos*, Blucher, São Paulo, 2012.

63. L. C. Basso, H. V. de Amorim, A. J. de Oliveira and M. L. Lopes, *FEMS Yeast Res.*, 2008, **8**, 1155–1163.

64. L. C. Basso, T. O. Basso and S. N. Rocha, in *Biofuel Production – Recent Developments and Prospects*, ed. M. A. D. S. Bernardes, InTech, 2011.

65. A. Pessoa and B. V. Kilikian, *Purificação de produtos biotecnológicos*, Manole, Barueri, 2005.

66. J. Fan, Y. C. Fan, Y. C. Pei, K. Wu, J. J. Wang and M. H. Fan, *Sep. Purif. Technol.*, 2008, **61**, 324–331.

67. P. G. Mazzola, A. M. Lopes, F. A. Hasmann, A. F. Jozala, T. C. V. Penna, P. O. Magalhaes, C. O. Rangel-Yagui and A. Pessoa, Jr., *J. Chem. Technol. Biotechnol.*, 2008, **83**, 143–157.

68. R. M. Banik, A. Santhiagu, B. Kanari, C. Sabarinath and S. N. Upadhyay, *World J. Microbiol. Biotechnol.*, 2003, **19**, 337–348.

69. A. D. G. Zuñiga, J. A. M. Pereira, J. S. R. Coimbra, L. A. Minim and E. E. G. Rojas, *Boletim Centro de Pesquisa de Processamento de Alimentos*, 2005, **2005**, 61–82.

70. J. S. R. Coimbra, F. Mojola and A. J. A. Meirelles, *J. Chem. Eng. Jpn.*, 1998, **31**, 277–280.

71. R. E. Treybal, *Mass-Transfer Operations*, 3rd edition, McGraw-Hill, New York, 1980.

72. M. J. Sebastião, J. M. S. Cabral and M. R. Aires-Barros, *J. Chromatogr. A*, 1994, **668**, 139–144.

73. P. G. Mazzola, A. M. Lopes, F. A. Hasmann, A. F. Jozala, T. C. V. Penna, P. O. Magalhaes, C. O. Rangel-Yagui and A. Pessoa Jr, *J. Chem. Technol. Biotechnol.*, 2008, **83**, 143–157.

74. L. F. P. Ferreira, M. E. Taqueda, M. Vitolo, A. Converti and A. Pessoa Jr, *J. Biotechnol.*, 2005, **116**, 411–416.

75. P. A. Albertsson, *Partition of Cell Particles and Macromolecules*, Wiley-Interscience, New York, 1986.

76. J. Møller-Jensen and J. Löwe, *Curr. Opin. Cell Biol.*, 2005, **17**, 75–81.

77. B.-J. Ni, B. E. Rittmann and H.-Q. Yu, *Trends Biotechnol.*, 2011, **29** 454–463.

78. P. B. Agrawal and A. B. Pandit, *Biochem. Eng. J.*, 2003, **15**, 37–45.

79. A. P. J. Middelberg, *Biotechnol. Adv.*, 1995, **13**, 491–551.

80. H. Schutte and M. R. Kula, *Biotechnol. Appl. Biochem.*, 1990, **12**, 599–620.
81. D. Liu, X.-A. Zeng, D.-W. Sun and Z. Han, *Innovative Food Sci. Emerging Technol.*, 2013, **18**, 132–137.
82. A. R. Kleinig and A. P. J. Middelberg, *Chem. Eng. Sci.*, 1998, **53**, 891–898.
83. W. Kelly and K. Muske, *Bioprocess Biosyst. Eng.*, 2004, **27**, 25–37.
84. R. Halim, T. W. T. Rupasinghe, D. L. Tull and P. A. Webley, *Bioresour. Technol.*, 2013, **140**, 53–63.
85. Y. Chisti and M. Moo-Young, *Enzyme Microbial Technol.*, 1986, **8**, 194–204.
86. B. Kunasundari and K. Sudesh, *Express Polym. Lett.*, 2011, **5**, 620–634.
87. N. Jacquel, C. W. Lo, Y. H. Wei, H. S. Wu and S. S. Wang, *Biochem. Eng. J.*, 2008, **39**, 15–27.
88. J. A. Ramsay, E. Berger, R. Voyer, C. Chavarie and B. A. Ramsay, *Biotechnol. Techniques*, 1994, **8**, 589–594.
89. A. Neves and J. Müller, *Biotechnol. Prog.*, 2012, **28**, 1575–1580.
90. L. M. Monks, A. Rigo, M. A. Mazutti, J. Vladimir Oliveira and E. Valduga, *Biocatal. Agric. Biotechnol.*, 2013, **2**, 165–169.
91. J. A. Asenjo, B. A. Andrews and J. M. Pitts, *Ann. NY Acad. Sci.*, 1988, **542**, 140–152.
92. B. A. Andrews and J. A. Asenjo, *Trends Biotechnol.*, 1987, **5**, 273–277.
93. T. R. Hopkins, *Bioprocess Technol.*, 1991, **12**, 57–83.
94. T. Vicente, J. P. B. Mota, C. Peixoto, P. M. Alves and M. J. T. Carrondo, *Biotechnol. Adv.*, 2011, **29**, 869–878.
95. A. M. Azevedo, P. A. J. Rosa, I. F. Ferreira and M. R. Aires-Barros, *Trends Biotechnol.*, 2009, **27**, 240–247.
96. G. Carta and A. Jungbauer, *Protein Chromatography: Process Development and Scale-Up*, Wiley-VCH, Weinheim, 2010.
97. U. Gottschalk, in *Disposable Bioreactors*, eds. R. Eibl and D. Eibl, 2009, vol. 115, pp. 171–183.
98. K. Kallberg, H. O. Johansson and L. Bulow, *Biotechnol. J.*, 2012, 7, 1485–1495.
99. W. Kopaciewicz and F. E. Regnier, *Anal. Biochem.*, 1983, **133**, 251–259.
100. A. A. Shukla, S. S. Bae, J. A. Moore and S. M. Cramer, *J. Chromatogr. A*, 1998, **827**, 295–310.
101. C. Borrmann, C. Helling, M. Lohrmann, S. Sommerfeld and J. Strube, *Sep. Sci. Technol.*, 2011, **46**, 1289–1305.
102. L. R. Pereira, D. M. F. Prazeres and M. Mateus, *J. Sep. Sci.*, 2010, **33**, 1175–1184.
103. X. D. Geng and L. L. Wang, *J. Chromatogr. B: Biomed. Appl.*, 2008, **866**, 133–153.
104. M. M. Diogo, J. A. Queiroz and D. M. F. Prazeres, *Bioseparation*, 2001, **10**, 211–220.
105. R. Valdes, B. Reyes, T. Alvarez, J. Garcia, J. A. Montero, A. Figueroa, L. Gomez, S. Padilla, D. Geada, M. C. Abrahantes, L. Dorta, D. Fernandez, O. Mendoza, N. Ramirez, M. Rodriguez, M. Pujol, C. Borroto and J. Brito, *Biochem. Biophys. Res. Commun.*, 2003, **310**, 742–747.

106. R. Morenweiser, *Gene Therapy*, 2005, **12**, S103–S110.
107. F. Sousa, D. M. F. Prazeres and J. A. Queiroz, *Trends Biotechnol.*, 2008, **26**, 518–525.
108. S. Mori and H. G. Barth, *Size Exclusion Chromatography*, Springer-Verlag, New York, 1999.
109. C. J. Fee, *Biotechnol. Bioeng.*, 2003, **82**, 200–206.
110. G. S. Chaga, *J. Biochem. Biophys. Methods*, 2001, **49**, 313–334.
111. S. Loughran and D. Walls, in *Protein Chromatography*, eds. D. Walls and S. T. Loughran, Humana Press, 2011, vol. 681, pp. 311–335.
112. T. Yang, M. C. Sundling, A. S. Freed, C. M. Breneman and S. M. Cramer, *Anal. Chem.*, 2007, **79**, 8927–8939.
113. J. Thommes and M. Etzel, *Biotechnol. Prog.*, 2007, **23**, 42–45.
114. J. Kostal, A. Mulchandani and W. Chen, *Biotechnol. Bioeng.*, 2004, **85**, 293–297.
115. A. Eon-Duval, K. Gumbs and C. Ellett, *Biotechnol. Bioeng.*, 2003, **83** 544–553.
116. L. Maranga, P. Rueda, A. F. G. Antonis, C. Vela, J. P. M. Langeveld, J. I. Casal and M. J. T. Carrondo, *Appl. Microbiol. Biotechnol.*, 2002, **59**, 45–50.
117. V. M. M. Goncalves, M. Takagi, R. B. Lima, H. Massaldi, R. C. Giordano and M. M. Tanizaki, *Biotechnol. Appl. Biochem.*, 2003, **37**, 283–287.
118. A. Nilsson, H. O. Johansson, M. Mannesse, M. R. Egmond and F. Tjerneld, *Biochim. Biophys. Acta Proteins Proteomics*, 2002, **1601**, 138–148.
119. S. Fexby, A. Nilsson, G. Hambraeus, F. Tjerneld and L. Bülow, *Biotechnol. Prog.*, 2004, **20**, 793–798.
120. A. Kumar, M. Kamihira and B. Mattiasson, in *Methods for Affinity-Based Separations of Enzymes and Proteins*, ed. M. N. Gupta, Birkhaeuser Verlag, Basel, 2002, pp. 163–180.
121. D. T. Kamei, J. A. King, D. I. C. Wang and D. Blankschtein, *Biotechnol. Bioeng.*, 2002, **78**, 203–216.
122. H. S. Choonia and S. S. Lele, *Sep. Purif. Technol.*, 2013, **110**, 44–50.
123. F. Y. Gu, J. Gao, J. Z. Xiao, Q. H. Chen, H. Ruan and G. Q. He, *Romanian Biotechnol. Lett.*, 2012, **17**, 7853–7862.
124. V. V. Kumar, M. P. Premkumar, V. K. Sathyaselvabala, S. Dineshkirupha, J. Nandagopal and S. Sivanesan, *Eng. Life Sci.*, 2011, **11**, 607–614.
125. S. Rajeeva and S. S. Lele, *Biochem. Eng. J.*, 2011, 54.
126. S. Rawdkuen, A. Vanabun and S. Benjakul, *Process Biochem.*, 2012, **47**, 2566–2569.
127. S. M. Harde and R. S. Singhal, *Sep. Purif. Technol.*, 2012, **96**, 20–25.
128. T. M. Przybycien, N. S. Pujar and L. M. Steele, *Curr. Opin. Biotechnol.*, 2004, **15**, 469–478.
129. V. K. Garg, S. E. Zale, A. R. M. Azad and O. D. Holton, *Role of Functionalized Membrane Separation Technology in Downstream Processing of Biotechnology-Derived Proteins*, edited by P. S. Todd, K. Sikdar and M. Bier, American Chemical Society Conference Proceedings Series:

Frontiers In Bioprocessing II, Boulder, Colorado, USA, American Chemical Society, Washington, DC, 1992.

130. A. S. Rathore and A. Shirke, *Prep. Biochem. Biotechnol.*, 2011, **41**, 398–421.
131. V. Teplyakov, E. Sostina, I. Beckman and A. Netrusov, *World J. Microbiol. Biotechnol.*, 1996, **12**, 477–485.
132. A. Strancar, M. Barut, A. Podgornik, P. Koselj, H. Schwinn, P. Raspor and D. Josic, *J. Chromatogr. A*, 1997, **760**, 117–123.
133. A. Maruška, in *J. Chromatography Library*, eds. T. B. T. František Švec and D. Zdeněk, Elsevier, 2003, vol. 67, pp. 143–172.
134. Y.-W. Wang, Q. Wu and G.-Q. Chen, *Biomaterials*, 2004, **25**, 669–675.
135. A. Maehara, S. Taguchi, T. Nishiyama, T. Yamane and Y. Doi, *J. Bacteriol.*, 2002, **184**, 3992–4002.
136. N. Jacquel, C.-W. Lo, Y.-H. Wei, H.-S. Wu and S. S. Wang, *Biochem. Eng. J.*, 2008, **39**, 15–27.
137. R. Rai, D. M. Yunos, A. R. Boccaccini, J. C. Knowles, I. A. Barker, S. M. Howdle, G. D. Tredwell, T. Keshavarz and I. Roy, *Biomacromolecules*, 2011, **12**, 2126–2136.
138. P. Furrer, M. Schmid, A. Hinz, E. Pletscher, S. Panke and M. Zinn, *Eur. Cells Mater.*, 2004, **7**, 30–31.
139. S. F. W. Sherborn, M. Alkington, T. G. Cambridge and D. M. H. Somerville, *US Pat.* 6,245,537, Removing endotoxin with an oxidizing agent from polyhydroxyalkanoates produced by fermentation, 2001.
140. L. T. Lim, R. Auras and M. Rubino, *Prog. Polym. Sci.*, 2008, **33**, 820–852.
141. M. S. Lopes, A. L. Jardini and R. M. Filho, *Proc. Eng.*, 2012, **42**, 1402–1413.
142. B. Gupta, N. Revagade and J. Hilborn, *Prog. Polym. Sci.*, 2007, **32**, 455–482.
143. C. Miller, A. Fosmer, B. Rush, T. McMullin, D. Beacom and P. Suominen, in *Comprehensive Biotechnology*, 2nd edition, ed. M.-Y. Murray, Academic Press, Burlington, 2011, pp. 179–188.
144. K. M. Nampoothiri, N. R. Nair and R. P. John, *Bioresour. Technol.*, 2010, **101**, 8493–8501.
145. L. Avérous, in *Monomers, Polymers and Composites from Renewable Resources*, eds. B. Mohamed Naceur and G. Alessandro, Elsevier, Amsterdam, 2008, pp. 433–450.
146. M. A. Abdel-Rahman, Y. Tashiro and K. Sonomoto, *Biotechnol. Adv.*, 2013, **31**, 877–902.
147. R. Datta and M. Henry, *J. Chem. Technol. Biotechnol.*, 2006, **81**, 1119–1129.
148. Y.-J. Wee, J.-S. Yun, Y. Y. Lee, A.-P. Zeng and H.-W. Ryu, *J. Biosci. Bioeng.*, 2005, **99**, 104–108.
149. A. Lyddiatt and D. A. O'Sullivan, *Curr. Opin. Biotechnol.*, 1998, **9**, 177–185.
150. G. N. M. Ferreira, G. A. Monteiro, D. M. F. Prazeres and J. M. S. Cabral, *Trends Biotechnol.*, 2000, **18**, 380–388.
151. M. Li and Y. X. Qiu, *Vaccine*, 2013, **31**, 1264–1267.
152. M. C. M. Mellado, J. A. Mena, A. Lopes, O. T. Ramirez, M. J. T. Carrondo, L. A. Palomares and P. M. Alves, *Biotechnol. Bioeng.*, 2009, **104**, 674–686.

CHAPTER 3

Enhanced Biomass Degradation by Polysaccharide Monooxygenases

TANGHE MAGALI,*[a] DANNEELS BARBARA,[a] STALS INGEBORG[b] AND DESMET TOM[a]

[a] Centre for Industrial Biotechnology and Biocatalysis, Faculty of Bioscience Engineering, Ghent University, Coupure Links 653, B-9000 Ghent, Belgium; [b] Department of Biochemistry, Faculty of Applied Engineering Sciences, University College Ghent, Schoonmeersstraat 52, B-9000 Ghent, Belgium
*Email: magali.tanghe@ugent.be

3.1 Introduction

'Lignocellulosic biomass' is a broad term for wood, agricultural residues, municipal solid waste and dedicated energy crops that contain a significant fraction of structural polysaccharides.[1] As cellulose is the most abundant biological material on earth, but cannot be digested by humans, it is the preferred renewable resource for bio-based industries.[2-4] Unfortunately, (lingo)cellulose is very recalcitrant towards enzymatic degradation, which has been a serious hurdle for the development of cost-effective bio-refineries.[5] There are three major reasons for this. First, in the plant cell wall, cellulose is embedded in a matrix of hemicellulose and lignin, limiting its accessibility and making pre-treatment steps crucial.[1] Second, cellulose chains are associated in so-called microfibrils, which have a high degree of

RSC Green Chemistry No. 27
Renewable Resources for Biorefineries
Edited by Carol Sze Ki Lin and Rafael Luque

hydrogen bonding and are insoluble in water and most other solvents. And finally, the β-1,4-glycosidic bond in cellulose is considerably more stable than its α counterpart present in starch, resulting in an uncatalysed half-life time of several million years.[6,7]

Until very recently, the enzymatic degradation of cellulose was thought to be brought about by hydrolytic enzymes only, *i.e.* a mixture of endoglucanases and cellobiohydrolases.[3] However, it was always hard to imagine how an individual cellulose chain could get separated from a microfibril and brought into the active site of these cellulases. Therefore, the presence of a 'missing link' in the process was already suggested in 1950, although it remained undetected for many years.[8] With the recent discovery of polysaccharide monooxygenases (PMO), a major step towards better understanding of biomass degradation has been taken. This novel class of fungal metalloproteins, belonging to the Auxiliary Activity family 9 (AA9, previously GH-61), degrades cellulose in a manner completely different to that of cellulases and is able to process crystalline substrates very efficiently.[9,10] Hence, PMOs render cellulose more susceptible to attack by the canonical cellulases and exploitation of their properties should result in the development of more efficient enzyme cocktails for low-cost lignocellulosic biomass conversion.[9,11]

In this chapter, an overview will be given of the current knowledge about this intriguing new class of enzymes. We will start by explaining the synergetic relationship between PMOs and cellulases in more detail. Next, we will discuss the structure of the enzymes, which gives clues about their physiological function. Then, their mode of action and reaction mechanism will be described, illustrating a very subtle yet powerful use of copper as metal cofactor. To end this chapter, available assays for activity measurements will be compared and a conclusion will be drawn.

3.2 Synergy in Cellulose Degradation

An overview of cellulose degradation is presented in Figure 3.1, providing a general framework to explain the role of PMOs. The classical view describes the collaboration of three types of enzymes, *i.e.* endo- and exo-acting cellulases as well as β-glucosidases.[12] The first are also called endo-1,4-β-glucanases (EG, EC 3.2.1.4) and cut the chain at internal positions. In this way, they provide new accessible ends for the exo-1,4-β-glucanases or cellobiohydrolases (CBH, EC 3.2.1.91). These enzymes are known to act upon the end of cellulose chains, releasing cellobiose as product. Finally, the oligosaccharides generated by the EGs and CBHs are converted to single glucose molecules by β-glucosidases (BG, EC 3.2.1.21). The majority of cellulases contain a carbohydrate-binding module (CBM), which helps them stick to their crystalline substrate. Nevertheless, they still prefer to work on soluble substrates since these can enter their active-site cleft or tunnel much more easily.

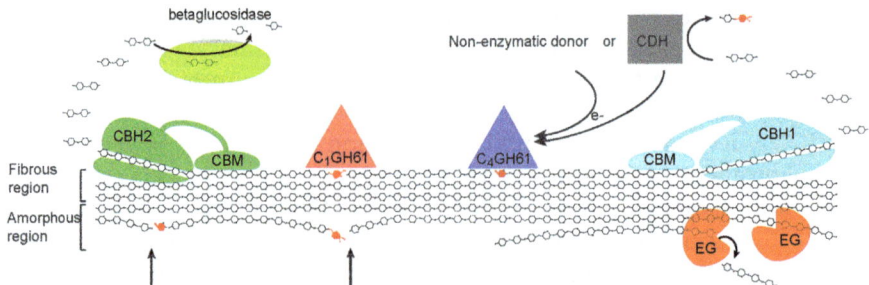

Figure 3.1   The synergetic mechanism of different classical cellulases and PMOs
(indicated as C1 and C4 GH-61). CBH: cellobiohydrolase, EG: endoglucanase, CDH: cellobiose dehydrogenase, CBM: carbohydrate binding
module.
Figure from Horn *et al.*[12]

Breaking up cellulose fibrils into soluble chains requires a substantial
amount of energy, which means that the pre-treatment of biomass is a
crucial (yet expensive) step. However, PMOs are able to bind directly to
crystalline cellulose thanks to a flat active site that is located on the outside of the protein's structure.[13] By cutting the chains and introducing
oxygen atoms (see below), they can disrupt the substrate's crystalline
structure and thus facilitate the action of cellulases. Furthermore, the
electrons needed for this oxidative cleavage can be delivered by redox active compounds like gallic acid or ascorbic acid or the enzyme cellobiose
dehydrogenase (CDH, EC 1.1.99.18), that acts on the main hydrolysis
product and thus provides a positive feedback loop for this cooperative
system. Consequently, a high degree of synergy is to be expected between
the redox enzymes on the one hand, and the hydrolytic enzymes on the
other hand.

To illustrate this point, the addition of PMO to a commercial cellulose
mixture was found to significantly increase the degree of hydrolysis of cellulosic substrates.[14] Furthermore, the enhancement was clearly proportional
to the amount of PMO added.[15] The individual cellulose specificities that are
present in the mixture were also evaluated in combination with PMO, where
it could be concluded that PMOs enhance the work of cellulases in a generic
way rather than improving the action of one specific type of cellulose.
Overall, adding PMOs to the mixture could almost halve the protein loading
required for a specific degree of hydrolysis.[12,13] Since the price of the enzymes contributes quite significantly to the cost of biomass saccharification,
the optimization of the enzyme mixture will be a key factor in making
lignocellulosic biomass a suitable and economically efficient renewable
resource. Furthermore, the need for biomass pre-treatment might also be
diminished since PMOs can process insoluble crystalline substrates and
make them more accessible for hydrolysis.

3.3 Protein Structure

The first PMO structure that was resolved is that of Cel61B from *Trichoderma reesei*.[13] By now, five other PMO structures have been studied, *i.e.* four from ascomycete fungi (*Thermoascus aurantiacus*,[9] *Thielavia terrestris*[14] and *Neurospora crassa*[11]) and more recently one from the basidiomycete fungus *Phanerochaete chrysosporium*.[16] Their overall fold will be discussed first, after which a close-up of their active site will be presented.

3.3.1 Overall Fold

The location of the active site in PMOs is very different from that in canonical cellulases. Indeed, the latter contain a binding cleft or tunnel, whereas the former have a wedge-like structure with a flat binding surface that can adhere to crystalline substrates. Their fold comprises two antiparallel and twisted β-sheets that form a fibronectin-III-like structure (Figure 3.2).[14] Most PMOs are composed of a single domain, although about 20% do include an additional carbohydrate binding module 1 (CBM1) to bring the catalytic domain in close contact with the substrate.[17] CBM1 is known to specifically bind cellulose and chitin, with three aromatic residues forming a flat binding surface to associate with these polysaccharide chains.[16]

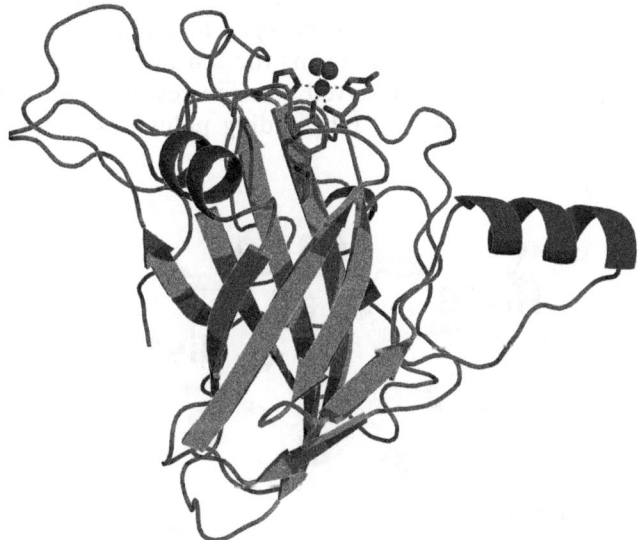

Figure 3.2 The first resolved PMO structure, Cel61B from the fungus *Trichoderma reesei*. A wedge-like structure can be observed where the residues in the active site are depicted in the flat binding surface. A detailed view of the active site is shown in Figure 3.3.

Since PMOs are secreted, post-translational modifications can occur. For example, *T. reesei* Cel61B contains two disulfide bridges, one methylation and one *N*-glycosylation site. In fact, the highly conserved histidine at the N-terminus is mostly methylated at its Nε2 position.[9] The reason is still unknown, although it doesn't seem to be essential for activity.[16] The *N*-glycosylation site in PMO-3 from *Neurospora crassa* is found close to the active site, forming an extension of the flat binding surface. The appearance of an alpha 3$_{10}$-helix in the second and highly variable loop seems to be correlated to this glycosylation and seems to direct the active site towards the substrate. The same combination was found in Cel61B, but was absent in other family members.[11]

3.3.2 Conservation

3.3.2.1 The Hexacoordination Centre as Active Site

The active site of a PMO holds a divalent copper ion that is required for the activity of the protein. This divalent cation is bound in a type-2 copper binding site, a structure that is also referred to as a 'hexacoordination centre' since six keystones are important in the binding (Figure 3.3). At the N-terminus, the first residue of the mature protein is a highly conserved histidine. This is one of the rare examples in nature where the N-terminus is important for protein activity. The histidine residue has a dual binding function: the first binding is made by the N-terminal amine, while the second interaction is made with the nitrogen at the δ-position. A second highly conserved histidine residue forms the 3th binding point, while the 4th binding point is a less conserved tyrosine residue or alternatively a proline. And finally, the 5th and 6th binding points are water molecules

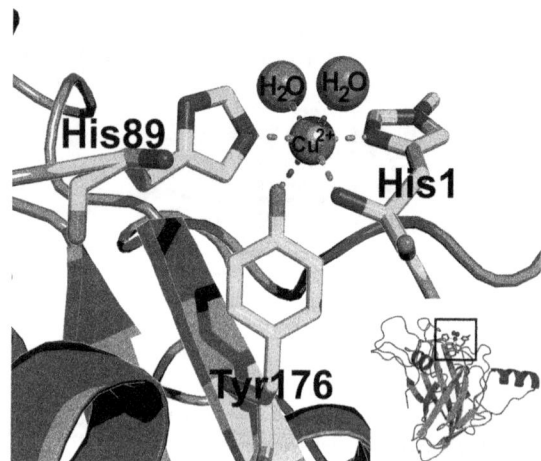

Figure 3.3 Active site of *Trichoderma reesei* Cel61B. The box in the bottom right corner indicates the location on the overall fold of the protein.

(which can be replaced by other molecules including peroxide, oxygen or sulfate ions). This coordination centre is the most conserved region in the enzyme and forms the active site. Since two histidine residues are important for the binding of the metal ion, Quinlan and co-workers referred to this architecture as a 'histidine brace'. A highly similar structure has also been found in methane monooxygenases.[9]

3.3.2.2 Other Putative Interaction Regions

Other conserved regions are less distinct and their functionality is still under investigation. Generally, the number of sheets in the β-sandwich can vary between different PMOs. The loops that interconnect these sheets differ in sequence, length and secondary structure between different members of family AA9 (formerly GH-61). It was suggested that these loops, especially the ones in extension to the binding surface, fulfill a role in molecular recognition.[11] The highly variable L2 loop contains tyrosine residues at the surface that are ideally spaced for binding glucose units in cellulose.[16] Another conserved region is a hydrophilic patch at the surface of the wedge, which is thought to allow association with its physiological redox partner cellobiose dehydrogenase (CDH). In that way, that enzyme could provide the electrons that are required for the reaction *via* a conserved hydrogen network that leads from CDH to the active site of the PMO.[11]

3.4 The Oxidative Reaction

When a PMO acts on cellulose, it generates a native and an oxidized polysaccharide. More insight in the different reaction components and the mechanism of action will be given in this section.

3.4.1 Reaction Components

3.4.1.1 Substrate: Cellulose

PMOs are supposed to be active only on (semi)crystalline polysaccharides and not on short soluble oligosaccharides (DP 2-5).[18] Harris *et al.* tried a whole variety of substrates and concluded that mixed substrates like PCS (pre-treated corn stover) generate higher activities than pure cellulose substrates like Avicel (microcrystalline cellulose) or PASC (phosphoric acid swollen cellulose).[14,15] However, hemicellulose substrates do not tend to be more efficiently degraded by PMO addition.[10] The first successful experiments with real pre-treated biomass, namely pre-treated spruce, were recently performed.[16]

3.4.1.2 Divalent Metal Ion

PMO activity is dependent on the presence of a divalent metal ion in the active site. After some initial confusion, an extremely high copper binding affinity has been shown by several research groups.[9,16,19,20] Addition of this metal ion to the reaction mixture is not required since only catalytic amounts are needed, which are already included during enzyme production.[18]

3.4.1.3 Electron Donor

The reaction of PMO needs to be supplied with electrons from a donor such as CDH. In addition, reducing agents such as ascorbic acid, reduced glutathione or gallic acid can be used in a concentration of about 1 mM.[9,18] Also lignin or hemicellulose substrates were suggested as enhancers due to their capacity of quenching free radicals.[15,21]

3.4.1.4 Molecular Oxygen

Molecular oxygen is consumed during the reaction, as demonstrated by $^{18}O_2$ experiments.[22] Previous studies indicated that O_2 mimics like sodium azide or cyanide can bind in the active site and lower the cellulose conversion.[10,20]

3.4.2 Location of Oxidation

Evidence is increasing that the PMOs can be classified in different groups according to the place of oxidation in a glucose unit of cellulose. The first experiments revealed the oxidation of the C1 atom in the scissile bond, generating an aldonic acid and a neutral saccharide. This is probably the most common oxidation pattern, but certainly not the only one. Indeed, Beeson *et al.* provided evidence for C4 oxidized chains by the cleavage of *Neurospora crassa* PMO, resulting in a 4-keto-aldose as product.[22] Furthermore, Langston and co-workers were able to identify C6 oxidized products, thus involving a position that is not even part of the glycosidic bond.[10,23]

3.4.3 Putative Reaction Mechanism

The first insights into the reaction mechanism of this interesting class of enzymes were provided by Vaaje-Kolstad and co-workers in 2008 in their work dealing with the related class Auxiliary Activity family 10 (AA10, see next paragraph). They suggested a mechanism that was a combination of a hydrolytic and an oxidative step. Two oxygen atoms are involved in this reaction, namely one from a water molecule and the other from molecular oxygen. The same mechanism was suggested for PMOs (Figure 3.4).[24] As a matter of fact, it was proven that one oxygen atom from molecular oxygen is built in, demonstrating its function as mono-oxygenase.[22]

Two years later, a more elaborate reaction mechanism was proposed by Li and co-workers.[11] This mechanism involved reduction and oxidation of copper on the enzyme itself and binding of oxygen with formation of several radicals, but ending in a C1, C4 or C6 oxidized glucose molecule. Supporting evidence for this mechanism was obtained later. In the highly similar

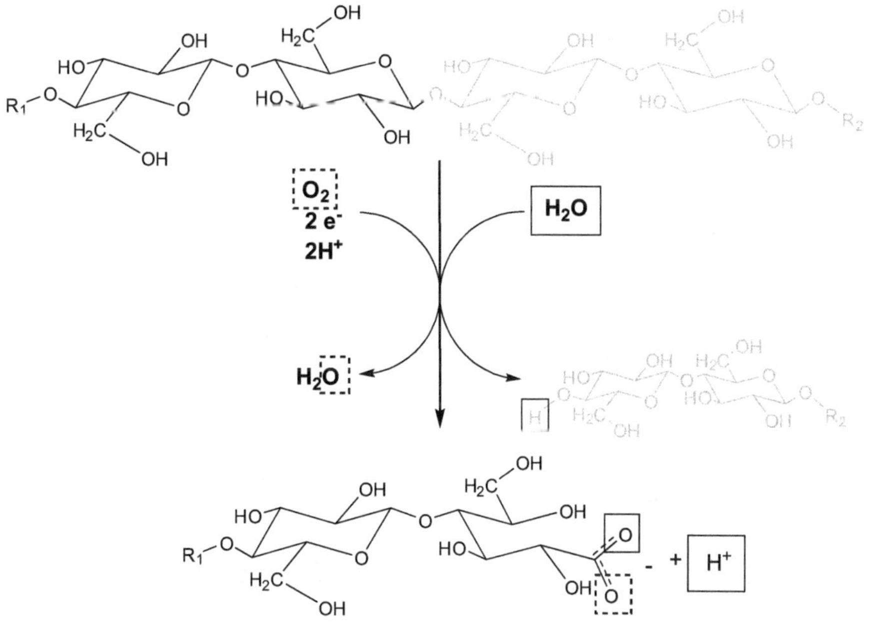

Figure 3.4 Reaction mechanism for C1-oxidation of a glucose in cellulose. One oxygen in the carboxyl group originates from molecular oxygen, while the other oxygen comes from a water molecule. Two electrons and two hydrogens are consumed in the reaction.

enzymes from family AA10, Cu^{1+} seems to bind with higher affinity than Cu^{2+}, suggesting that copper is reduced in the enzyme itself.[20] Such results were also reported for PMOs by Philips *et al.* in 2011.[25]

3.4.4 Similar Reactions

The described putative role and reaction mechanism of PMOs was assigned based on similarity to proteins that are categorized as Auxiliary Activity family 10 (AA10, formerly CBM33).[4] Enzymes from family AA10 are secreted mainly by bacteria, but also by a variety of other organisms like insects, fungi and viruses. In 2005, it was for the first time discovered that an enzyme from family AA10, namely CBP21 (chitin binding protein 21), facilitates the action of hydrolytic enzymes by increasing the substrate accessibility.[26] Indeed, AA10 proteins introduce oxygen atoms in the chitin chain, thereby opening up the crystalline structure and rendering the substrate more accessible for chitinases. The main resemblances and differences between members from family AA9 and AA10 are described below.

First, their substrate and cooperative degradation process is highly similar. As a matter of fact, chitin and cellulose are crystalline analogues (Figure 3.5) that are degraded by a similar collaboration of endo- and exo-acting enzymes. Secondly, the three-dimensional structure of PMOs

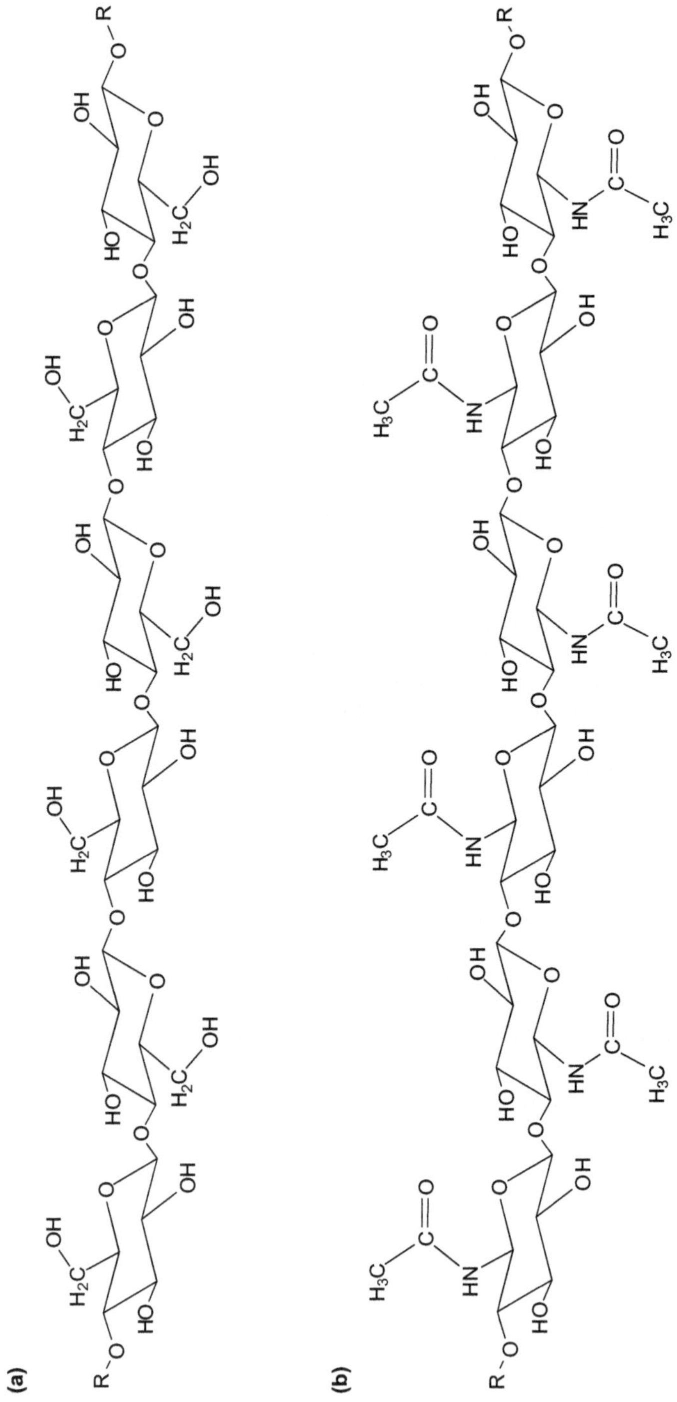

Figure 3.5 Cellulose (panel A) and chitin (panel B) are structural analogues.

resembles that of enzymes from family AA10, although their active sites are slightly different: the latter have 4 metal coordination points instead of 6, and their terminal histidine is never methylated. Thirdly, the cofactors involved in the reaction are identical for both enzymes, as a reductant and a divalent metal ion are always indispensable. And finally, the reaction products of chitin treated with AA10 are similar to the reaction products of cellulose treated with PMO, namely an oxidized and a neutral sugar chain. Moreover, some members of AA10 were found to enhance not only chitin degradation by chitinases but also cellulose degradation by cellulases. Examples of such enzymes are CelS2 secreted by *Streptomyces coelicolor* A3(2)[27] and two AA10-containing proteins secreted by *Thermobifida fusca*.[28]

However, a crucial difference is that the oligosaccharides formed by CBP21 are even numbered, while the oligosaccharides from cellulose can also have odd numbers of glucose. This tendency might be caused by the use of less crystalline substrates for cellulose degradation experiments so that the enzyme can attack from any side. In contrast, with real biomass such as spruce, a trend towards even numbered reaction products seems to be present.[16]

3.5 Activity Measurement

Measuring the activity of PMOs is a very challenging task. Not only the need for insoluble substrates but also the multiplicity of reaction products and the high tendency of chains with introduced oxygen to remain attached to the crystalline substrate bedevils the activity tests.

3.5.1 Colorimetric Methods

Colorimetric assays attempt to measure the reaction progress in a fast and easy way without revealing every single component in the product mixtures. The first steps have been taken in this challenging domain, but it still remains in its infancy.

Activity measurement of classical cellulases can be performed by measuring reducing sugars (*e.g.* BCA or DNS), but for this oxidizing enzyme such methods cannot be used, as was reported by Harris *et al.*[14] However, they can be used to reflect the synergetic action of the PMO together with the classical cellulases.

A method that was introduced by Kittl and co-workers in 2012 measures a side reaction of a PMO that seems of no further importance in the cellulose degradation process. In the absence of cellulose especially, the enzyme reduces O_2 to H_2O_2 (Figure 3.6). Coupling of this reaction to the reduction of amplex red to resorufin in assistance of horseradish peroxidase (HRP), results in an assay where the formation of the fluorescent resorufin is proportional to the quantity of PMO added. The only electron donors tested in this reaction were cellobiose dehydrogenase (CDH) with addition of lactose to be reduced and ascorbate. The reaction is selective and sensitive to the

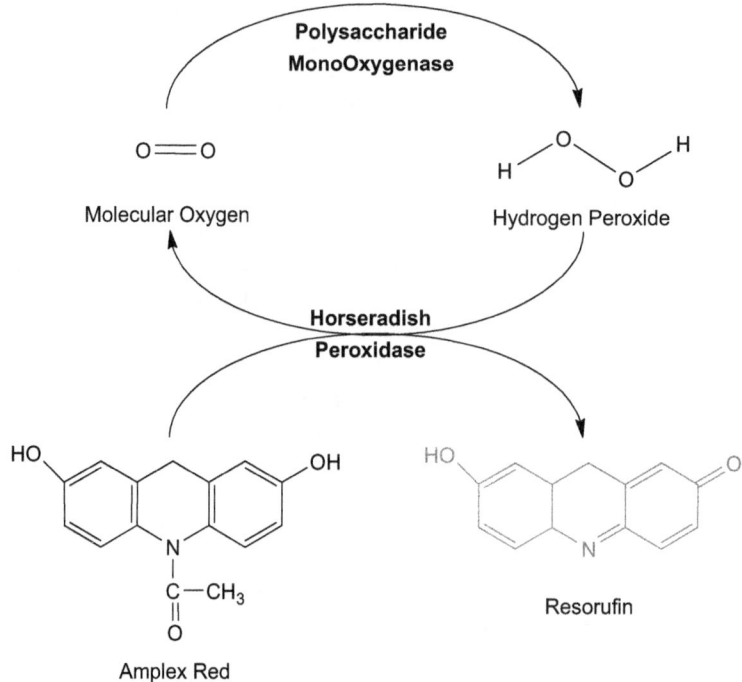

Figure 3.6 Colorimetric measurement method for PMO activity based on a side reaction that is performed by the PMO and reduces molecular oxygen to hydrogen peroxide (developed by Kittl *et al.*, 2012). The coupled reaction which oxidizes amplex red to the fluorescent molecule resorufin is responsible for fluorescence formation.

operation of PMOs. Nevertheless care should be taken to applying this reaction since it only is a side reaction.

3.5.2 Analytical Methods

Lack of a fast method has obliged researchers so far to use more complex methods. The advantage is that one gets much more information about the different products formed in the product mixture. This is interesting in unravelling the reaction mechanism. However, identification remains a thorny issue considering the fact that a huge amount of products are formed: oxidized versus non-oxidized and oxidation at different positions and all possible degrees of polymerization. The production of all standards is usually impossible. Besides, these methods are more time consuming.

3.5.2.1 HPLC

High-performance liquid chromatography (HPLC) is a technique regularly used in analytical chemistry to separate the components of a mixture.

A mobile phase with varying composition is pushed through a column under high pressure, while different components are separated by their difference in absorption force to the packaging of the column. As detector in PMO activity experiments, a refractive index (RI) detector is mostly coupled to a regular HPLC system.[9,10,14]

A more specific method that is regularly used in oligosaccharide detection is high-performance anion exchange chromatography (HPAEC) which applies a high pH to ionize the sugars. The anion exchange column is usually combined with a pulsed amperometric detection (PAD) system that oxidizes the sugars at a gold electrode. This system is highly applicable for the separation of carbohydrates. HPAEC-PAD is extensively used in PMO research, in which the elution time takes about 1 hour for each sample.[11,18,22,23,27] More chromatographic techniques were compared for the separation of native and oxidized oligosaccharides by Westereng *et al.*[29] They concluded that out of high-performance anion-exchange chromatography (HPAEC), hydrophilic interaction chromatography (HILIC) and porous graphitized carbon liquid chromatography (PGC-LC), the HPAEC-PAD system was the best method in separation oligosaccharides with DP 1-10. They shortened the method to less than 10 minutes so that the method is applicable for high-throughput screening.[29]

3.5.2.2 PACE

PACE stands for polysaccharide analysis by using carbohydrate gel electrophoresis. This method is the analogue of using an agarose gel to separate DNA fragments, but is specifically designed for carbohydrate separation.[30] A drawback is that an additional step is required for derivatization of the product by a colouring agent like ANTS (8-aminonapthalene-1,3,6-trisulphonic acid). Quinlan and co-workers used this technique to evaluate the necessity of copper in the reaction of PMOs.[9]

3.5.2.3 Mass Spectrometry

Identification of different reaction products or different peaks from chromatograms is done by using standards or performing mass spectrometry. MALDI-TOF is an ionization technique frequently used.[9,18,23,27] Also, combined setups can be found like LC-MS[22] or UPLC-ESI-MS (electron spray ionization mass spectrometry) profiles.[10]

3.6 Classification

Polysaccharide monooxygenases have long been classified as Glycoside Hydrolase family 61 (GH-61) because weak endoglucanase activity had been observed. These experimental results might be explained by the fact that chain breaks introduced near the reducing end will release soluble native cello-oligosaccharides. Nevertheless, this classification is incorrect since an oxidative rather than a hydrolytic cleavage is performed. Since then, other and more accurate names have been proposed, such as copper metallo-enzyme, oxidohydrolases, (lytic) polysaccharide monooxygenase and so on. In 2013, these enzymes were reclassified as Auxiliary Activity family 9 (AA9) in the CAZy database (Carbohydrate Active enZymes, www.cazy.org).[31]

Interestingly, several organisms produce multiple PMO enzymes that display a high sequence variability. For example, the soil fungus *Haetomium globosum* contains 44 unique PMO genes.[12] Even more remarkable, some fungi have more PMOs than cellulases. *Neurospora crassa* for example, codes for 10 PMOs with an average pairwise sequence identity of 33%. Most likely, these variants have a different specificity for cellulose and attack different shapes and crystal forms of the substrate.[11] Furthermore, it has been suggested by some authors that different PMOs are required together to attack cellulose under different angles.[12] It can be concluded that PMOs play an underestimated role in biomass degradation, but a detailed understanding of their mode of action is still lacking.[11] Anyhow, researchers are trying to find a classification system to organize these variable sequences in subclasses. A system was proposed, based on the location of oxidation: PMO type-I include the PMOs that introduce a C1 oxidation, PMO type-II include a C4 oxidation and PMO type-III are less specific. However, not all PMOs seem to fit in this system.[23] Another system was suggested by Busk and Lange, who divided the AA9 family based on short conserved sequences using an algorithm called Peptide Pattern Recognition. The algorithm led to 16 subfamilies where the already characterized PMOs belong to only six of these subclasses. The 16 subfamilies thus probably are an overestimation, as was also the case for other families analysed in this way.[32]

3.7   Conclusion

PMOs are believed to facilitate the action of canonical cellulases by loosening the crystal packing of their substrate through an oxidative cleavage mechanism. Our knowledge about these enzymes is gradually increasing but is still far from complete. Nevertheless, their application potential is extremely high because of their crucial role in the conversion of lignocellulosic biomass as a renewable resource. Although the first steps in unravelling the mechanism of this new enzyme class have already been taken, a number of important questions still need to be answered. Some are listed below:

1. Why is there so much multiplicity within the class of PMOs? And why is the multiplicity within the related AA10 family so much smaller?[18]
2. Which physiological partner delivers the reducing power required for the action of PMOs? Some cellulolytic organisms do not produce a cellobiose dehydrogenase, so other factors must be able to take over that role.[27]
3. Does the addition of PMO lowers the need for biomass pre-treatment? If so, the whole process of biomass conversion might need to be re-evaluated and re-optimized.

Acknowledgements

The authors wish to thank the Institute for the Promotion of Innovation through Science and Technology in Flanders (IWT-Vlaanderen) for financial support through a PhD grant.

References

1. K. A. Gray, L. Zhao and M. Emptage, *Curr. Opin. Chem. Biol.*, 2006, **10**, 141–146.
2. V. Menon and M. Rao, *Prog. Energy Combustion Sci.*, 2012, **38**, 522–550.
3. L. R. Lynd, P. J. Weimer, W. H. van Zyl and I. S. Pretorius, *Microbiol. Mol. Biol. Rev.*, 2002, **66**, 506–577.
4. S. T. Merino and J. Cherry, *Adv. Biochem. Eng. Biotechnol.*, 2007, **108** 95–120.
5. D. Klein-Marcuschamer, P. Oleskowicz-Popiel, B. A. Simmons and H. W. Blanch, *Biotechnol. Bioeng.*, 2012, **109**, 1083–1087.
6. D. B. Wilson, *Curr. Opin. Microbiol.*, 2011, **14**, 259–263.
7. R. Wolfenden, X. Lu and G. Young, *J. Am. Chem. Soc.*, 1998, **120**, 6814–6815.
8. E. T. Reese, R. G. H. Siu and H. S. Levinson, *J. Bacteriol.*, 1950, **59** 485–497.
9. R. J. Quinlan, M. D. Sweeney, L. Lo Leggio, H. Otten, J. C. Poulsen, K. S. Johansen, K. B. Krogh, C. I. Jorgensen, M. Tovborg, A. Anthonsen, T. Tryfona, C. P. Walter, P. Dupree, F. Xu, G. J. Davies and P. H. Walton, *Proc. Natl. Acad. Sci. USA*, 2011, **108**, 15079–15084.
10. J. A. Langston, T. Shaghasi, E. Abbate, F. Xu, E. Vlasenko and M. D. Sweeney, *Appl. Environ. Microbiol.*, 2011, **77**, 7007–7015.
11. X. Li, W. T. T. Beeson, C. M. Phillips, M. A. Marletta and J. H. Cate, *Structure*, 2012, **20**, 1051–1061.
12. S. J. Horn, G. Vaaje-Kolstad, B. Westereng and V. G. Eijsink, *Biotechnol. Biofuels*, 2012, **5**, 45.
13. S. Karkehabadi, H. Hansson, S. Kim, K. Piens, C. Mitchinson and M. Sandgren, *J. Mol. Biol.*, 2008, **383**, 144–154.
14. P. V. Harris, D. Welner, K. C. McFarland, E. Re, J.-C. Navarro Poulsen, K. Brown, R. Salbo, H. Ding, E. Vlasenko, S. Merino, F. Xu, J. Cherry, S. Larsen and L. L. Leggio, *Biochemistry*, 2010, **49**, 3305–3316.
15. M. Dimarogona, E. Topakas, L. Olsson and P. Christakopoulos, *Bioresour. Technol.*, 2012, **110**, 480–487.
16. M. Wu, G. T. Beckham, A. M. Larsson, T. Ishida, S. Kim, C. M. Payne, M. E. Himmel, M. F. Crowley, S. J. Horn, B. Westereng, K. Igarashi, M. Samejima, J. Stahlberg, V. G. Eijsink and M. Sandgren, *J. Biol. Chem.*, 2013, **288**, 12828–12839.
17. T. T. Teeri, A. Koivula, M. Linder, G. Wohlfahrt, C. Divne and T. A. Jones, *Biochem. Soc. Trans.*, 1998, **26**, 173–177.

18. B. Westereng, T. Ishida, G. Vaaje-Kolstad, M. Wu, V. G. Eijsink, K. Igarashi, M. Samejima, J. Stahlberg, S. J. Horn and M. Sandgren, *PloS one*, 2011, **6**, e27807.

19. R. Kittl, D. Kracher, D. Burgstaller, D. Haltrich and R. Ludwig, *Biotechnol. Biofuels*, 2012, **5**, 79.

20. F. L. Aachmann, M. Sorlie, G. Skjak-Braek, V. G. Eijsink and G. Vaaje-Kolstad, *Proc. Natl. Acad. Sci. USA*, 2012, **109**, 18779–18784.

21. D. Cannella, C. W. Hsieh, C. Felby and H. Jorgensen, *Biotechnol. Biofuels*, 2012, **5**, 26.

22. W. T. Beeson, C. M. Phillips, J. H. Cate and M. A. Marletta, *J. Am. Chem. Soc.*, 2012, **134**, 890–892.

23. M. Bey, S. Zhou, L. Poidevin, B. Henrissat, P. M. Coutinho, J. G. Berrin and J. C. Sigoillot, *Appl. Environ. Microbiol.*, 2013, **79**, 488–496.

24. G. Vaaje-Kolstad, B. Westereng, S. J. Horn, Z. Liu, H. Zhai, M. Sorlie and V. G. Eijsink, *Science*, 2010, **330**, 219–222.

25. C. M. Phillips, W. T. Beeson, J. H. Cate and M. A. Marletta, *ACS Chem. Biol.*, 2011, **6**, 1399–1406.

26. G. Vaaje-Kolstad, S. J. Horn, D. M. van Aalten, B. Synstad and V. G. Eijsink, *J. Biol. Chem.*, 2005, **280**, 28492–28497.

27. Z. Forsberg, G. Vaaje-Kolstad, B. Westereng, A. C. Bunaes, Y. Stenstrom, A. MacKenzie, M. Sorlie, S. J. Horn and V. G. Eijsink, *Protein Sci.*, 2011, **20**, 1479–1483.

28. F. Moser, D. Irwin, S. Chen and D. B. Wilson, *Biotechnol. Bioeng.*, 2008, **100**, 1066–1077.

29. B. Westereng, J. W. Agger, S. J. Horn, G. Vaaje-Kolstad, F. L. Aachmann, Y. H. Stenstrom and V. G. Eijsink, *J. Chromatogr. A*, 2013, **1271**, 144–152.

30. F. Goubet, P. Jackson, M. J. Deery and P. Dupree, *Anal. Biochem.*, 2002, **300**, 53–68.

31. A. Levasseur, E. Drula, V. Lombard, P. M. Coutinho and B. Henrissat, *Biotechnol. Biofuels*, 2013, **6**, 41.

32. P. K. Busk and L. Lange, *Appl. Environ. Microbiol.*, 2013, **79**, 3380–3391.

CHAPTER 4

Microalgae Technology

ZHENG SUN, YAN-HUI BI AND ZHI-GANG ZHOU*

College of Fisheries and Life Science, Shanghai Ocean University,
Shanghai 201306, People's Republic of China
*Email: zgzhou@shou.edu.cn

4.1 Introduction

Microalgae represent a diverse group of eukaryotic photosynthetic micro-organisms that are either unicellular or multicellular in form. It is estimated that there are over 50 000 species present in nature. They can survive under a wide range of environmental conditions, from water to land and even in snow and desert soils.[1]

Microalgae are a huge source of valuable natural ingredients, and they have numerous potential applications in many industrial fields. It is note-worthy that in recent years, the exploitation of microalgae as the feedstock for biofuel production has attracted particular attention globally. Re-searchers believe that microalgae possess significant advantages over traditional biofuel feedstocks such as plant oil and animal fats. As sunlight-driven cell factories converting CO_2 to biomass, microalgae exhibit much higher photosynthetic efficiency than land plants. As a result, most species double themselves within one day, and the fastest-growing species *Chloro-cuccum littorale* has a doubling time as short as 8 hours.[2] High oil content is also an important feature of microalgae. In general, microalgae synthesize a relatively low level of oil under favorable growth conditions, but the oil content can be significantly enhanced by nutrient deficiency, in particular nitrogen. So far, the highest oil content is 73% of dry weight, achieved in a *Scenedesmus* sp. strain with nutrient starvation for 11 days.[3] Another

RSC Green Chemistry No. 27
Renewable Resources for Biorefineries
Edited by Carol Sze Ki Lin and Rafael Luque
© The Royal Society of Chemistry 2014
Published by the Royal Society of Chemistry, www.rsc.org

advantage of microalgae is that they cause no competition with food for arable lands. If soybean, a popular oil crop in US is used for biofuel production, 5.2 times the area of US cropland will need to be employed to meet existing transport fuel needs.[4,5] The huge requirement for arable lands, as well as the resultant conflicts between food and oil, can be resolved by microalgae. As microalgae can be successfully grown in a broad range of environmental conditions, much less land is needed if they are used as biodiesel feedstock: only 0.7% the area needed for soybean.[4,5] These unique properties make microalgae the most promising cell factory for biofuel production, with significant potential to displace fossil diesel.[4] Many studies have proved that the microalgal biofuel is of good quality. The key properties, *e.g.* energy density, viscosity, flash point, cold filter plugging point and acid value, comply with the specifications established by American Society for Testing and Materials.[6] In this chapter, the current status of microalgae-based biofuel is introduced, including cultivation technologies, production pipeline, engineering approaches for strain improvement, as well as the biorefinery-based integrated production of oil and co-products.

4.2   Mass Cultivation

Microalgae biomass is produced in specially engineered facilities. Currently, open pond and photo bio-reactor (PBR) systems are the two most viable options for large-scale commercial production of algal biomass for biofuel.

4.2.1   Open Pond System

Open systems include raceway ponds, circular ponds and tanks. They resemble mostly closely the nature of microalgae and serve as the oldest and simplest systems for algal production. The simplicity lowers the production costs. An open pond usually is shallow (0.25–0.4 m deep) because optical absorption and self-shading by the algal cells limits light penetration through the algal broth.[4]

Although open systems are of low-cost and durable with large production capacity, they have substantial intrinsic disadvantages. It is difficult to control the cultivation environment in open ponds so that the algae are susceptible to contamination by other microorganisms. Also, the utilization of light by microalgae cells in an open pond is poor, which leads to a low cell density and biomass productivity. Other major drawbacks include the rapid water loss due to evaporation and the large amount of land occupied.

4.2.2   Photo Bio-reactor (PBR) System

The above mentioned limitations associated with open systems can be overcame by PBRs. PBRs refer to those closed systems in which all growth elements are introduced into the reactor and controlled according to the requirements. Compared with open ponds, PBRs are more flexible and require less space. Made by transparent materials with a large surface area-to-volume ratio, PBRs offer maximum efficiency in using light so that the biomass productivity can be greatly improved.[7] Also, contamination is effectively prevented under the controlled cultivation environment, which makes axenic algal monocultures possible.

In PBR systems, the control of oxygen accumulation could be a problem. The oxygen generated by photosynthesis will accumulate in the enclosed space and inhibit the algal growth. Thus, the culture must be periodically retuned to a degassing zone where the algal broth is bubbled with air to remove the excess oxygen.[4] In addition, steps like cooling, mixing and CO_2 feeding are all required, which makes PBRs much more expensive to build and operate than ponds. In this context, a hybrid system has been proposed. The system couples PBRs and open ponds in a two-stage process, using PBRs to produce contaminant-free inoculants which are subsequently applied in large open ponds. The estimated oil production cost is at $ 84 bbl^{-1} ('bbl' is a measuring unit referring to a barrel).[8]

In addition to the cost, another significant limitation of PBRs is the emission of CO_2. In the EU renewable energy directive, the CO_2 emission associated with algal biomass production is estimated by multiplying the external energy inputs to the process by the default emissions factors (Directive 2009/28/EC). It was found that in the course of algal biomass production, PBRs may give rise to greater carbon emissions than the conventional fossil diesel, whereas the emissions caused by open ponds could be comparable with rape methyl ester biodiesel. This is mainly associated with the electricity consumption for pumping and mixing and the provision of heat to dry the algae.[9] To overcome this problem, using co-products to generate electricity could be a promising strategy.

The comparison between open ponds and PBRs for microalgae cultivation is illustrated in Table 4.1.

4.3 Cultivation Mode

Microalgae are usually cultivated under photoautotrophic conditions. This culture mode is easy to maintain and thus has been commonly employed in microalgal industries. However, due to the hostile or unsteady environment, a high growth rate is hardly achieved. Besides, in photoautotrophic cultures, the light is often deficient because of the mutual shading of cells (especially in denser cultures), resulting in poor cell yield.[10] To solve the problems, another culture system, namely heterotrophic cultivation, has been suggested. The most remarkable feature of heterotrophic cultivation is the utilization of organic carbon substances as the sole carbon and energy source. Since the requirement for light is eliminated, a significant increase of cell density is possible, which leads to a much higher productivity than

Table 4.1 Comparison of open ponds and PBRs for microalgae cultivation.[5]

	Open ponds	PBRs
Contamination control	difficult	easy
Contamination risk	high	reduced
Sterility	none	achievable
Process control	difficult	easy
Species control	difficult	easy
Mixing	very poor	uniform
Operation regime	batch or semi-continuous	batch or semi-continuous
Area/volume ratio	low	high
Algal cell density	low	high
Investment	low	high
Operation cost	low	high
Light utilization efficiency	low	high
Temperature control	difficult	uniform
Productivity	low	high
Evaporation of growth medium	high	low
Gas transfer control	low	high
Oxygen inhibition	less severe than PBRs	great problem
Scale-up	difficult	difficult

photoautotrophic cultivation.[11] Heterotrophic cultivation can be carried out in bioreactors, similar to procedures established with bacteria or yeast.[12] Glucose is the most superior carbon source. It has been found that the ATP generated from heterotrophic mode using glucose as the energy supplier exceeds more than 600% over the photoautotrophic mode in which energy is supplied by light.[13] In contrast to light conditions where Embden-Meyerhof-Parnas (EMP) pathway acts as the predominant glycolysis, under darkness, glucose is mainly metabolized *via* the pentose phosphate (PP) pathway.[14] For example, in alga *Chlorella pyrenoidosa*, the PP pathway accounts for over 90% of glucose metabolic flux distribution.[13] So far, the maximum biomass content in heterotrophic cultures achieves up to 150 g L^{-1} of cell dry weight,[15] much higher than photoautotrophic cultures (<40 g L^{-1}).

Despite the benefits like high cell density and productivity, heterotrophic cultivation has its own disadvantages. First, the number of microalgal species to grow heterotrophically is very limited because only a small portion can survive in complete darkness. For most obligate photoautotrophic algae, they may lack the efficient uptake of essential substrates (especially sugars) into cells,[16] or the inside tricarboxylic acid (TCA) cycle is incomplete due to the absence of key enzymes such as α-ketoglutarate dehydrogenase.[17] Second, using glucose as the carbon source could be expensive. Thus, heterotrophic cultivation is preferably employed for the production of high value added ingredients rather than algal oil.

Table 4.2 shows the comparison between photoautotrophic and heterotrophic production of microalgal oils.

Table 4.2 Comparison between photoautotrophic and heterotrophic production of microalgal oils.

	Photoautotrophic mode	Heterotrophic mode
Available algal species	a lot	limited
Cell density	low	high
Cost of production	low	high
Fatty acid contents	low	high
Lipid contents	low	high
Productivity	low	high
Scale-up	easy	difficult

4.4 Biomass Processing

The conversion of microalgae to fuels takes the following steps: harvesting, dewatering, extraction and oil transesterification.

4.4.1 Harvesting

Efficient harvesting is essential for industrial production because microalgal cultures are relatively dilute. In open ponds, the cell density is usually less than 1 g L^{-1}. Harvesting of biomass is a two-stage process: (1) biomass is separated from the bulk suspension; and (2) the slurry is concentrated into a thick algal paste.[18] Most algal cells are negatively charged so that the formation of aggregates is prevented. Besides, microalgae have tiny cell dimensions and a similar specific gravity to water, which leads to a very slow sinking rate. These properties make algal suspension highly stable and difficult to harvest.

Major harvesting techniques include flocculation, flotation, gravity sedimentation, centrifugation and filtration.

Flocculation is a preparatory step prior to others. As algal cells are negatively charged, the addition of flocculants such as multivalent cations and cationic polymers can neutralize the surface charge. As a result, the particle size can be effectively increased to disrupt the stability of the system and facilitate the aggregation. Ferric chloride $(FeCl_3)$, aluminum sulfate $(Al_2(SO_4)_3)$ and ferric sulfate $(Fe_2(SO_4)_3)$ are commonly used multivalent metal salts.[18] The flocculation technique can be used to handle large quantities of culture, and it is less energy intensive than mechanical separation. However, this method alone is not sufficient for harvesting, and other processes need to be combined.

Unlike flocculation, flotation does not require any additional chemicals. Air or gas bubbles are pumped into the algae culture. Bubbles collapse near algal cells, and then get attached to the cells. As a result, the density of algal cells is reduced and they float. This method can capture particles with a diameter of less than 500 µm.[19] There are three major flotation techniques based on the bubble size, namely dissolved air flotation, dispersed air flotation and electrolytic flotation.

Filtration operated under pressure or vacuum is feasible for harvesting large microalgae (>70 μm) such as *Coelastrum*. To recover the smaller microalgae (<30 μm), membrane-based microfiltration and ultrafiltration may be applied,[20] but they are more expensive.

Similar to filtration, gravity sedimentation is also suitable for recovering the large microalgae. This method is usually considered as the first plan in algal wastewater treatment systems because of the large volumes treated and the low value of the biomass obtained.[21]

Centrifugation is an accelerated sedimentation process. It is considered as the most efficient harvesting technique, and can recover most microalgae from the liquid broth. In the laboratory test where pond effluent was set at 500–1000 g, about 80–90% of microalgae can be recovered within 2–5 min.[19] However, the exposure of algal cells to high gravitational and shear forces may cause some structural damage. Also, centrifugation is a capital- and energy-intensive technique. It is more appropriate to be used for the recovery of high-value products from microalgae rather than biofuel.

4.4.2   Dewatering

After harvesting, the biomass slurry containing 5–15% of dry solid must be processed rapidly, or it may spoil within a few hours in a hot climate.[22] Sun drying is the oldest method that works well in low-humidity climates. However, due to the high water content of algal biomass, this method is not very effective for algal powder production. Other dewatering methods include direct/indirect heating, fluid bed, flash drying and microwave drying. Compared with sun drying, they are more efficient, but also more energy intensive. It is estimated that the removal of 1 kg of water requires over 800 kcal of energy. Thus, dewatering is considered as one of the main economic bottlenecks in the entire process.

4.4.3   Extraction

Algae have a tough exterior to protect internal lipids. With a high elasticity modulus, cell walls are difficult to crack. In most cases, the oil extraction from dried biomass is a two-step process including mechanical disruption and solvent extraction. Mechanical disruption includes pressing, bead milling and homogenization. When 100–200 g L^{-1} of biomass concentrations are used, bead milling is the most effective and economical way.[23] Ultrasonic- and microwave-assisted extractions can also be used to improve the efficiency.

After being mechanically disrupted, the algal cells are exposed to solvents. The Bligh and Dyer procedure, originally designed to extract lipids from fish tissue, has been used as a benchmark for comparison of solvent extraction techniques.[24] A wide range of organic solvents such as benzene, cyclohexane, hexane, acetone and chloroform have been proved to be effective in treating algal cells, of which hexane is the major solvent used. In some cases, the solvent extraction can be enhanced by using organic solvents at temperature and pressures above the boiling point, which is known as the accelerated solvent extraction.[25] The mechanical disruption–solvent extraction process allows the solvent to significantly penetrate the biomass and make the best physical contact with the lipid materials. As a result, more than 95% of the total oil present in algae can be successfully extracted. Some other extraction techniques include osmotic shock, enzyme extraction, supercritical CO_2 extraction and supercritical methanol extraction.

4.4.4 Oil Transesterification

The oil obtained from microalgae usually has a higher viscosity than diesel oil, so that it cannot be applied to an engine directly. To reduce the viscosity and increase the fluidity, transesterification, a reaction between triglycerides and a short-chain alcohol is needed. Transesterification is not a new process as it has been well studied in the treatment of the vegetable oil. In this process, large, branched triglycerides are transformed into smaller, straight chain molecules, which are similar in size to the molecules of the species present in diesel fuel.[26]

Transesterification utilizes 1 mol of triglycerides and 3 mol of alcohol to produce 1 mol of glycerol and 3 mol of fatty esters (Figure 4.1). Since the reaction is reversible, large excess of alcohol needs to be prepared. The complete process occurs stepwise with the first conversion of triglycerides to diglycerides and then to monoglycerides and finally to glycerol. Numerous alcohols can be used as the substrates, such as methanol, ethanol, propanol, butanol and amyl alcohol. Methanol is the most preferable one because of its low cost. The most abundant composition of microalgal oil transesterified with methanol is $C_{19}H_{36}O_2$, which has been suggested to meet the standard of biofuel.[6]

As shown in Figure 4.1, a catalyst is of great importance to the transesterification process. Chemicals (acid and alkali) and enzymes can be used as the catalysts. Most commonly used acids include sulfuric, sulfonic, phosphoric and hydrochloric acids. These strong acids are corrosive and the processes are slow.[27] Alkali-catalyzed transesterification has a better performance than acid, of which the reaction rate is approximately 4000 times faster.[28] Thus, alkali is preferred for industrial production of biofuel. Most

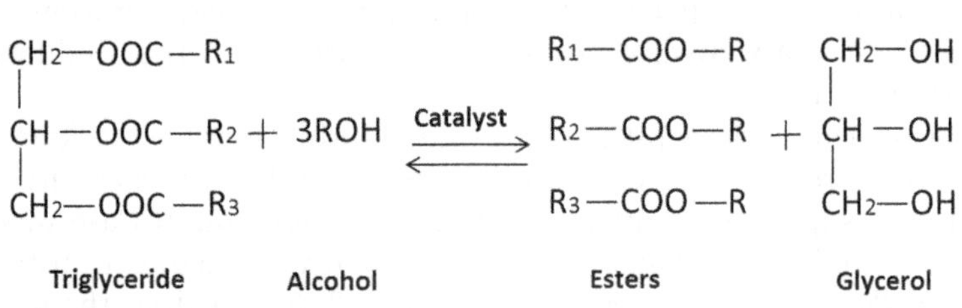

$$
\begin{array}{l}
CH_2\!-\!OOC\!-\!R_1 \\
| \\
CH\ -\!OOC\!-\!R_2\ +\ 3ROH \\
| \\
CH_2\!-\!OOC\!-\!R_3
\end{array}
\xrightarrow[\;\;\longleftarrow\;\;]{\text{Catalyst}}
\begin{array}{l}
R_1\!-\!COO\!-\!R \\
\\
R_2\!-\!COO\!-\!R\ +\ \\
\\
R_3\!-\!COO\!-\!R
\end{array}
\begin{array}{l}
CH_2\!-\!OH \\
| \\
CH\ -\!OH \\
| \\
CH_2\!-\!OH
\end{array}
$$

    Triglyceride   **Alcohol**   **Esters**   **Glycerol**

Figure 4.1 Transesterification of oil to biodiesel. R_{1-3} indicates hydrocarbon group.

commonly used types include sodium hydroxide (NaOH), potassium hydroxide (KOH) and sodium methoxide (CH_3ONa). Unlike the chemical catalysts that need to be removed by the end of transesterification, enzymes can ensure a high purity of the final product, so that the washing step can be omitted. Lipase is a well-studied biocatalyst.[27,29] However, the cost of enzyme is still relatively high and remains a barrier for its industrial implementation.

4.5 Improving the Economics of Microalgal Oil

Although microalgae show great potential as feedstock for biofuel production, microalgal oil is still far from economically viable due to the high production cost. The unit production cost of microalgal oil is estimated to be \$12.73 gal^{-1}, much higher than that of plant oil.[30] Substantial cost reduction is needed to make microalgal oil production competitive.

4.5.1 Genetic Engineering

An ideal algal species is expected to possess the desirable characteristics including rapid growth rate, high oil content, robustness, high CO_2 tolerance, large cell size, ease of disruption, and so on.[31] It is unlikely to find such a naturally occurring strain with all the above mentioned traits. Genetic engineering is considered as a potential approach towards enhancing microalgal biology for improved production economics of oil. Complete genome sequences of several microalgae are available now, including the red alga *Cyanidioschyzon merolae*,[32] the diatoms *Thalassiosira pseudonana*[33] and *Phaeodactylum tricornutum*,[34] and the green algae *Chlamydomonas reinhardtii*,[35,36] *Ostreococcus tauri*[37] and *Ostreococcus lucimarinus*.[38] Several microalgal genome sequencing projects are also ongoing. To date, more than 30 different strains have been successfully transformed.[39]

Neutral lipids, especially triacylglycerols (TAGs), are considered to be superior to other lipids for biofuel feedstock because of their higher content of fatty acid and lack of phosphate.[40] Thus, efforts have been taken to enhance the TAG content through genetic engineering. Several studies attempted to overexpress the genes involved in TAG assembly, such as glycerol-3-phosphate acyltransferase (GPAT),[41] lysophosphatidic acid acyltransferase (LPAAT)[42] and diacylglycerol acyltransferase (DGAT).[43] The findings are summarized in Table 4.3.

Acetyl-CoA carboxylase (ACCase), a rate-limiting enzyme, is known to catalyze the first committed step of fatty acid biosynthesis. Its overexpression is supposed to increase the lipid production. ACCase was successfully isolated from the diatom *Cyclotella cryptica*[45] and then transformed into *Cyclotella cryptica* and *Navicula saprophila*.[46] Nevertheless, the increase in lipid production was not observed. These results suggested that the ACCase alone might not be enough to regulate the whole pathway.

Table 4.3 Available information on algal acyltransferase involved in TAG biosynthesis.[42,44]

Gene source	Algae species	Host species	Validation parameter
GPAT	T. pseudonana	S. cerevisiae (gat 1 mutant)	Increased GPAT activity in yeast
	C. reinhardtii	S. cerevisiae (gat 1 mutant)	Increased TAG content in yeast Increased transcript level in algae
	P. incisa	Algae	Increased transcript level in algae Increased TAG content in algae
LPAAT	C. reinhardtii	N/A	N/A
DGAT	O. tauri	S. cerevisiae H1246 (TAG-deficient quadruple mutant)	Restoration of TAG biosynthesis Oil body formation in yeast
	T. pseudonana	S. cerevisiae H1246 (TAG-deficient quadruple mutant)	Restoration of TAG biosynthesis Oil body formation in yeast
	C. reinhardtii	C. reinhardtii (starchless mutant)	Increased transcript level in algae Increased TAG content in algae

In addition to the direct engineering of oil biosynthesis, oil enhancement can be achieved by the manipulation of transcriptional factors that regulate oil biosynthesis,[47,48] or by the blocking of competing metabolic pathways that share the common carbon precursors, *e.g.* starch biosynthesis.[49] Genetic engineering can also be adopted to alter fatty acid compositions of oil for biofuel quality improvement, *e.g.* heterologous expression of thioesterase genes to accumulate shorter chain length fatty acids[50] or inactivation of Δ12 desaturase gene to produce higher level of oleic acids at the cost of polyunsaturated fatty acids.[51]

4.5.2 Biorefinery-based Production Strategy

As mentioned at the beginning of this chapter, microalgal biomass contains not only oil but also a wide range of valuable ingredients, including carotenoids, polyunsaturated fatty acids, polymers, peptides, vitamins, enzymes, amino acids, and so on. Some of them exhibit significant market potential whilst some have already achieved commercial success. Table 4.4 shows the health benefits of some of the main bioactive compounds. From a biorefinery point of view, the residual biomass after oil extraction may be used as food additives, nutraceuticals and animal feed.

Table 4.4   Bioactive compounds produced by microalgae and their potential health benefits.

Products	Microalgae	Potential health benefits	Ref.
Borophycin	*Nostoc spongiaeforme* var. *tenue*	cancer prevention	52
Carotenoids astaxanthin beta-carotene canthaxanthin lutein zeaxanthin	*Chlorella* sp. *Dunaliella salina* *Haematococcus pluvialis* *Spirulina* sp.	antioxidant properties cancer prevention cardiovascular protection DNA damage protection eye and skin protection glycation inhibition immune system enhancement inflammation prevention neuroprotection	53, 54
Cyanovirin-N	*Nostoc ellipsosporum*	anti-Ebola virus anti-HCV/HIV/HSV virus anti-influenza virus	55
Fatty acids arachidonic acid docosahexaenoic acid eicosapentaenoic acid oleic acid	*Botryococcus braunii* *Chlorella* sp. *Crypthecodinium cohnii* *Nitzschia* sp. *Schizochytrium* sp.	cancer prevention cardiovascular protection immune system enhancement inflammation prevention mental and visual strengthening	56, 57
Lipopeptides	*Anabaena spiroides* *Microsystis aeruginosa* *Synechocystis trididemni*	antibiotic anti-cancer anti-virus enzyme inhibition	58
Phycobiliprotein allophycocyanin c-phycocyanin phycoerythrin	*Spirulina platensis* *Porphyridium* sp.	cancer prevention inflammation prevention anti-virus antioxidant properties neuroprotection	59, 60
Polysaccharides	*Chlorella* sp. *Nostoc* sp. *Porphyridium* sp.	antioxidant properties anti-virus immune system enhancement UV protection	61
Squalene	*Schizochytrium* sp. *Thraustochytrium* sp.	cancer prevention cardioprotection immune system enhancement skin protection	56
Vitamins tocopherols (Vitamin E)	*Porphyridium* sp. *Spirulina platensis*	antioxidant properties	61

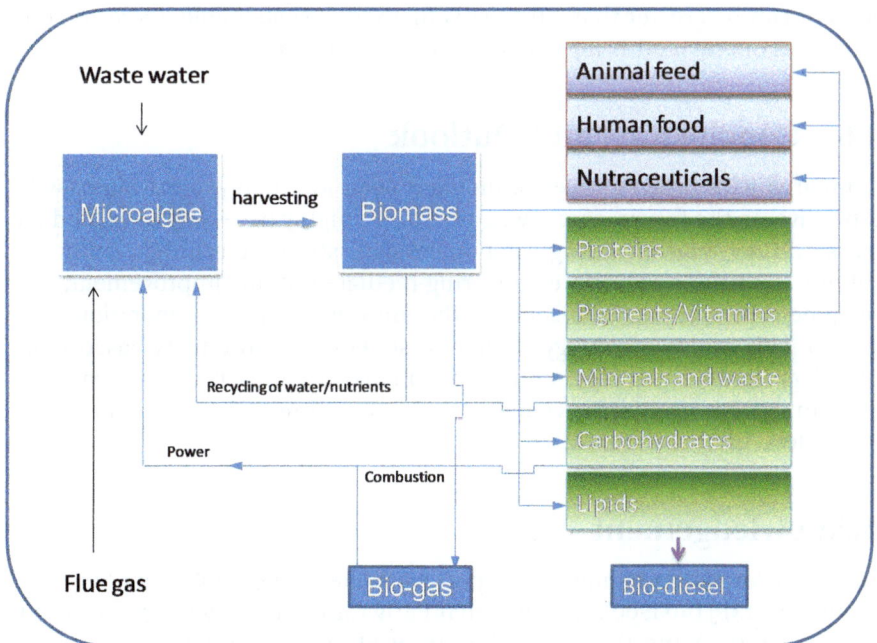

Figure 4.2 A schematic illustration of integrated production of microalgal oils and other products coupled with the possible flue gas and wastewater treatments.

For example, the green alga *Chlorella zofingiensis* is able to synthesize a red pigment, namely astaxanthin, in addition to accumulating oil inside its cells.[62] Astaxanthin is a carotenoid with potent antioxidant capacity. Its market demand is huge in nutraceutical, cosmetics, food and feed industries with an international price higher than \$3000 kg^{-1}. The integrated production of oil and high value-added production, coupled with the possible recycling of water and nutrients, is a highly promising strategy towards reducing the production cost of microalgal oil (Figure 4.2).

4.5.3 Development of Low-cost Carbon Source

For those algae which can survive in complete darkness, heterotrophic culture could be better than phototrophic culture as it exhibits much higher cell density and oil productivity. In this context, finding a cheap carbon sources is a highly promising strategy. To date, various by-products or wastes from industrial and agricultural processes have been evaluated as the potential feedstock, including Jerusalem artichoke,[63] sweet sorghum,[64] wheat bran,[65] the wastewater from sugar or milk processing industries,[66] crude glycerol,[67] waste molasses[62,68] as well as the food waste containing rice, noodles, meat and vegetables.[69] Some of these studies show promising results that the lipid content can be greater than 50%. This method is sustainable, increases the process economy and brings environmental benefits.

4.6   Conclusions and Outlook

Compared with other sources of biomass used for energy, algal biomass is expensive. In the future, cost-saving efforts are highly necessary to expand its market. They may include greater microalgal strain screening, optimized culture conditions, genetic engineering-mediated strain improvement, development of cheap carbon sources, the innovation of next generation PBR system with improved energy efficiency, state-of-art biorefinery-based integrated production strategy, and so on. Breakthroughs in these areas will determine the economic viability and environmental sustainability of microalgae biofuel industry.

Acknowledgement

Authors acknowledge financial support from the Shanghai Pujiang Program (Grant No. 13PJ1403500), the Shanghai Early Career Faculty Program (Grant No. ZZHY13005) and the Doctoral Fund of Ministry of Education of China (Grant No. 20133104120004).

References

1. R. E. Lee, *Phycology*, 4th edn, Cambridge University Press, Cambridge, 2008.
2. M. Ota, Y. Kato, H. Watanabe, M. Watanabe, Y. Sato, R. L. Smith, Jr. and H. Inomata, *Bioresour. Technol.*, 2009, **100**, 5237.
3. T. Matsunaga, M. Matsumoto, Y. Maeda, H. Sugiyama, R. Sato and T. Tanaka, *Biotechnol. Lett.*, 2009, **31**, 1367.
4. Y. Chisti, *Biotechnol. Adv.*, 2007, **25**, 294.
5. T. M. Mata, A. A. Martins and N. S. Caetano, *Renew. Sustain. Energy Rev.*, 2010, **14**, 217.
6. H. Xu, X. Miao and Q. Wu, *J. Biotechnol.*, 2006, **126**, 499.
7. C. U. Ugwu, H. Aoyagi and H. Uchiyama, *Bioresour. Technol.*, 2008, **99**, 4021.
8. M. E. Huntley and D. G. Redalje, *Mitig. Adapt. Strat. GL.*, 2007, **12**, 273.
9. R. Slade and A. Bauen, *Biomass Bioenerg.*, 2013, **53**, 29.
10. Z. Y. Wen and F. Chen, *Biotechnol. Adv.*, 2003, **21**, 273.
11. F. Chen, *Trends Biotechnol.*, 1996, **14**, 421.
12. F. Bumbak, S. Cook, V. Zachleder, S. Hauser and K. Kovar, *Appl. Microbiol. Biot.*, 2011, **91**, 31.
13. C. Yang, Q. Hua and K. Shimizu, *Biochem. Eng. J.*, 2000, **6**, 87.
14. O. Perez-Garcia, F. M. E. Escalante, L. E. de-Bashan and Y. Bashan, *Water Res.*, 2011, **45**, 11.

15. M. E. de Swaaf, J. T. Pronk and L. Sijtsma, *Biotechnol. Bioeng.*, 2003, **81**, 666.
16. R. A. Lewin, *Physiology and Biochemistry of Algae*, 1st edn, Academic Press, New York, 1962.
17. C. R. Benedict, *Annu .Rev. Plant Phys.*, 1978, **29**, 67.
18. L. Brennan and P. Owende, *Renew. Sust. Energ. Rev.*, 2010, **14**, 557.
19. C. Y. Chen, K. L. Yeh, R. Aisyah, D. J. Lee and J. S. Chang, *Bioresour. Technol.*, 2011, **102**, 71.
20. B. Petrusevski, G. Bolier, A. N. Van Breemen and G. J. Alaerts, *Water Res.*, 1995, **29**, 1419.
21. Y. Nurdogan and W. J. Oswald, *Water Sci. Technol.*, 1996, **33**, 229.
22. E. Molina Grima, E. H. Belarbi, F. G. Acien Fernandez, A. Robles Medina and Y. Chisti, *Biotechnol. Adv.*, 2003, **20**, 491.
23. H. C. Greenwell, L. M. Laurens, R. J. Shields, R. W. Lovitt and K. J. Flynn, *J. R. Soc. Interface*, 2010, 7, 703.
24. E. G. Bligh and W. J. Dyer, *Can. J. Biochem. Physiol.*, 1959, **37**, 7.
25. M. Cooney, G. Young and N. Nagle, *Sep. Purif. Rev.*, 2009, **38**, 291.
26. S. Sinha, A. K. Agarwal and S. Garg, *Energy Convers. Manage.*, 2008, **49**, 1248.
27. H. Taher, S. Al-Zuhair, A. H. Al-Marzouqi, Y. Haik and M. Farid, *Enzyme Res.*, 2011, **46**, 8292.
28. H. Fukuda, A. Kondo and H. Noda, *J. Biosci. Bioeng.*, 2009, **92**, 405.
29. L. Fjerbaek, K. V. Christensen and B. Norddahl, *Biotechnol. Bioeng.*, 2009, **102**, 1298.
30. J. W. Richardson, M. D. Johnson and J. L. Outlaw, *Algal Res.*, 2012, **1**, 93.
31. R. H. Wijffels and M. J. Barbosa, *Science*, 2010, **329**, 796.
32. M. Matsuzaki, O. Misumi, T. Shin-I, S. Maruyama, *et al.*, *Nature*, 2004, **428**, 653.
33. E. V. Armbrust, J. A. Berges, C. Bowler, B. R. Green, *et al.*, *Science*, 2004, **306**, 79.
34. C. Bowler, A. E. Allen, J. H. Badger, J. Grimwood, *et al.*, *Nature*, 2008, **456**, 239.
35. J. Shrager, C. Hauser, C. W. Chang, E. H. Harris, *et al.*, *Plant Physiol.*, 2003, **131**, 401.
36. S. S. Merchant, S. E. Prochnik, O. Vallon, E. H. Harris, *et al.*, *Science*, 2007, **318**, 245.
37. E. Derelle, C. Ferraz, S. Rombauts, P. Rouze, *et al.*, *Proc. Natl. Acad. Sci. USA*, 2006, **103**, 11647.
38. B. Palenik, J. Grimwood, A. Aerts, P. Rouze, *et al.*, *Proc. Natl. Acad. Sci. USA*, 2007, **104**, 7705.
39. R. Radakovits, R. E. Jinkerson and M. C. Darzins Al Posewitz, *Eukaryot. Cell*, 2010, **9**, 486.
40. J. Pruvost, G. Van Vooren, G. Cogne and J. Legrand, *Bioresour. Technol.*, 2009, **100**, 5988.
41. J. Xu, Z. Zheng and J. Zou, *Botany*, 2009, **87**, 544.

42. I. Khozin-Goldberg and Z. Cohen, *Biochimie*, 2011, **93**, 91.
43. M. Wagner, K. Hoppe, T. Czabany, M. Heilmann, *et al.*, *Plant Physiol. Biochem.*, 2010, **48**, 407.
44. G. Feng, L. Cheng, X. Xu, L. Zhang and H. Chen, *Prog. Chem.*, 2012, **24**, 1411.
45. P. G. Roessler, *Plant Physiol.*, 1990, **92**, 73.
46. T. G. Dunahay, E. E. Jarvis and P. G. Roessler, *J. Phycol.*, 1995, **31**, 1004.
47. N. M. D. Courchesne, A. Parisien, B. Wang and C. Q. Lan, *J. Biotechnol.*, 2009, **141**, 31.
48. N. R. Boyle, M. D. Page, B. Liu, I. K. Blaby, *et al.*, *J. Biol. Chem.*, 2012, **287**, 15811.
49. Y. Li, D. Han, G. Hu, M. Sommerfeld and Q. Hu, *Biotechnol. Bioeng.*, 2010, **107**, 258.
50. R. Radakovits, P. M. Eduafo and M. C. Posewitz, *Metab. Eng.*, 2011, **13**, 89.
51. G. Graef, B. J. Lavallee, P. Tenopir, M. Tat, *et al.*, *Plant Biotechnol. J.*, 2009, **7**, 411.
52. S. Singh, *Crit. Rev. Biotechnol.*, 2005, **25**, 73.
53. A. C. Guedes, H. M. Amaro and F. X. Malcata, *Mar. Drugs*, 2011, **9**, 625.
54. Z. Sun, X. Peng, J. Liu, W. F. Fan, M. Wang and F. Chen, *Food Chem.*, 2010, **120**, 261.
55. S. Xiong, J. Fan and K. Kitazato, *Appl. Microbiol. Biot.*, 2010, **86**, 805.
56. K. W. Fan, F. Chen, *Bioprocessing for Value-Added Products from Renewable Resources*, ed. S. T. Yang, Elsevier Science, 2007, pp. 293–323.
57. T. A. Mori, *Clin. Exp. Pharmacol. Physiol.*, 2006, **33**, 842.
58. A. M. Burja, B. Banaigs, E. Abou-Mansour, J. G. Burgess and P. C. Wright, *Tetrahedron*, 2001, **57**, 9347.
59. J. Huang, J. Wang, B. L. Chen, M. Z. Wang, *et al.*, *J. Plant. Res. Envrion.*, 2006, **15**, 20.
60. J. Riss, K. Decorde, T. Sutra, M. Delage, *et al.*, *J. Agric. Food Chem.*, 2007, **55**, 7962.
61. M. Plaza, M. Herrero, A. Cifuentes and E. Ibanez, *J. Agric. Food Chem.*, 2009, **57**, 7159.
62. J. Liu, J. Huang, Y. Jiang and F. Chen, *Bioresour. Technol.*, 2012, **107**, 393.
63. Y. Cheng, W. Zhou, C. Gao, K. Lan, Y. Gao and Q. Wu, *J. Chem. Technol. Biot.*, 2009, **84**, 777.
64. C. Gao, Y. Zhai, Y. Ding and Q. Wu, *Appl. Energ.*, 2010, **87**, 756.
65. M. M. EL-Sheekh, M. Y. Bedaiwy, M. E. Osman and M. M. Ismail, *Int. J. Recycl. Org. Waste Agr.*, 2012, **1**, 12.
66. F. Bumbak, S. Cook, V. Zachleder, S. Hauser and K. Kovar, *Appl. Microbiol. Biot.*, 2011, **91**, 31.
67. Y. H. Chen and T. H. Walker, *Biotechnol.Lett.*, 2011, **34**, 1.
68. D. Yan, Y. Lu, Y. F. Chen and Q. Wu, *Bioresour. Technol.*, 2011, **102**, 6487.
69. D. Pleissner, W. C. Lam, Z. Sun and C. S. K. Lin, *Bioresour. Technol.*, 2013, **137**, 139.

CHAPTER 5

Application of Food Waste Valorization Technology in Hong Kong

KWAN TSZ HIM,[a,b] CAROL SZE KI LIN[a] AND
CHAN KING MING*[b]

[a] School of Energy and Environment, City University of Hong Kong,
Hong Kong; [b] Environmental Science Program, School of Life Sciences,
Chinese University of Hong Kong, Hong Kong
*Email: kingchan@cuhk.edu.hk

5.1 Introduction

In 2018, it is estimated that all of the landfills in Hong Kong will be filled up, making it no longer an option for the disposal of waste.[1] With the consideration of the closure of landfill and the problems associated with the disposal of waste to landfills; food waste, which is potentially a renewable resource, should be diverted from landfills to valorization facilities in the foreseeable future.

5.1.1 Characteristics of Food Waste

'Food waste' can be described as any by-product or waste product from the production, processing, distribution and consumption of food.[2] It is characterized by its high proportion of organic matter and high moisture content. According to the literature reporting the characteristics of food waste,

RSC Green Chemistry No. 27
Renewable Resources for Biorefineries
Edited by Carol Sze Ki Lin and Rafael Luque
© The Royal Society of Chemistry 2014
Published by the Royal Society of Chemistry, www.rsc.org

their moisture content and carbon to nitrogen (C/N) ratios are 74–90% and 14.7–36.4, respectively.[3] In view of these characteristics of food waste, in cases where there is no special arrangement or treatment for food waste storage, the disposal of food waste is very difficult due to the growth of pathogens, the high moisture content and rapid autoxidation.[4] As there are already many different microorganisms in the food waste, the high rate of microbial activity encourages the growth of pathogens, which adds difficulty to the storage of food waste with the concern of the rise of odour, hygiene problems and infectious disease.[5] High moisture contents increase the cost of transportation and reduces the efficiency of dehydration process in some valorization technologies while the fat content in food waste is susceptible to rapid oxidation, which releases foul-smelling fatty acids adding difficulty to the storage and treatment of food waste.

5.1.2 Definitions of Food Waste

Nowadays, the definitions of food waste are varied in different countries or cities and the change in terminology often affects the regulations applied to food waste management as well. According to the United Nations' definitions, 'Food waste' and 'Food loss' are different. 'Food loss' refers the decrease in food quantity or quality which makes it unsuitable for human use[6] while 'food waste' refers to the food losses at the end of the food chain due to retailers' and consumers' behaviour.[7] In the US, the United States Environmental Protection Agency defines food waste as 'Uneaten food and food preparation wastes from residences and commercial establishments such as grocery stores, restaurants, and produce stands, institutional cafeterias and kitchens, and industrial sources like employee lunchrooms.'[8] In Hong Kong, there is no specific definition of food waste, while it generally refers to any wasted foods, raw materials before cooking and also edible materials from manufacturing, distribution and retail stages. Also, food waste is counted as a part of the municipal solid waste, which is the major waste in Hong Kong. On the other hand, the academic literature defines food waste with respect to where it is generated in the food supply chain (FSC), which consists of harvesting, threshing, distribution, storage, processing, packaging, retail and post-consumer stage in general.[7] Food wastes from different stages of the FSC are characterized in terms of the causes of food waste generation, the proportion of the total food waste generation in industrialized and developing countries and the potential of valorization.[7]

5.1.3 Limitations of the Application of Food Waste Valorization Technology

There are numerous options for food waste valorization technologies available around the world, such as composting, animal feed production, incineration for energy production and anaerobic digestion for biogas production. However, no single technology can eradicate the food waste problem from a city. Storage of food waste is a major obstacle of food waste valorization as food waste is susceptible to the growth of pathogens, causing health risks and hygiene problems. Then, in view of the lack of space for storage and the large amount of daily food waste generation, the food waste valorization facilities should be on a mega-scale size with enough treatment capacity to process numerous tonnes of food waste produced every day. It definitely requires a huge initial investment for setting up the large-scale facilities. Also, in case of off-site processing, the large volume and great weight of food waste significantly increases the transportation cost and time taken in collecting the food waste. Last but not least, the composition of food waste varies, affecting the quality and specification of the regenerated products. The regenerated products, such as compost and animal feed, have different composition or nutritive requirements. For example, the protein content of swine feed in Hong Kong is required to be above 20% and the price of the feed increases along with its protein content. Therefore, the varied composition of food waste makes it an unpromising source to produce high quality product and probably decreases the product's competiveness in the market.

5.2 Current Status of Food Waste in Hong Kong

In Hong Kong, food waste is classified into domestic food waste and commercial and industrial (C&I) food waste. These two kinds of food waste are different in terms of the composition, distribution of sources and quantities. First, the origin of the waste mirrors the heterogeneity of the food waste. Domestic food waste is mainly composed of the leftovers from the households while C&I food waste is mainly composed of the by-products and the food products from the food processing industry. Therefore, the composition of domestic food waste is varied and contains low nutritional value while the composition of C&I food waste is more homogeneous than domestic food waste and contains abundant functionalized molecules and valuable compounds. Also, the C&I food waste generation is more concentrated since food manufacturers generally centre their manufacturing activities on one product and it would therefore be easier to collect a large quantity of unified food waste from a single source while domestic food waste generation is mainly from separated sources due to the multi-storey and multi-tenant building setting in Hong Kong. In Hong Kong 88% households live in multi-tenant buildings of more than 10 storeys.[9] However, domestic food waste and C&I food waste are collected together and dumped in the landfill nowadays. For the quantity, the amount of food waste generation is staggering in recent years. According to the annual reports of 'Monitoring of Solid Waste in Hong Kong' released by the Environmental Protection Department, the amount of food waste generated daily has shown an increasing trend from 3155 tonnes in 2002 to 3584 tonnes in 2011 (Figure 5.1) and has become the major component in the municipal solid

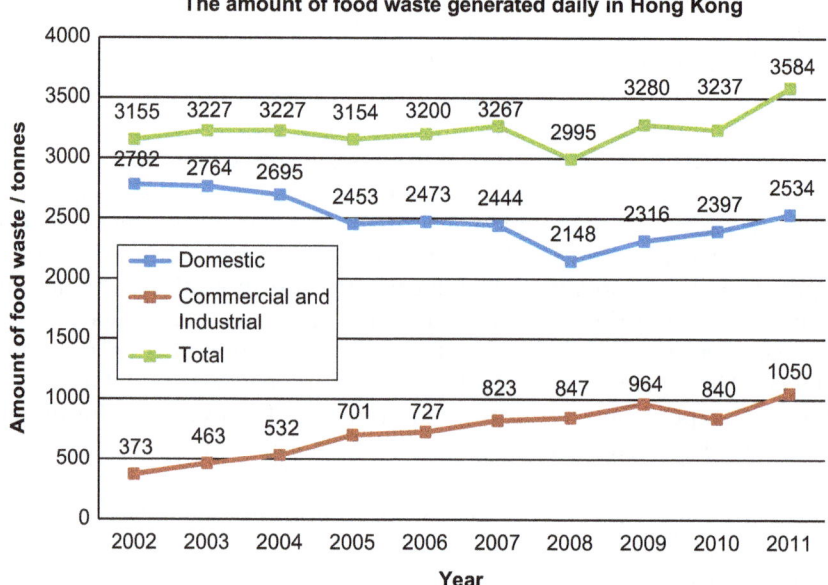

Figure 5.1 The amount of domestic and C&I food waste generated daily in Hong Kong from 2002 to 2011.

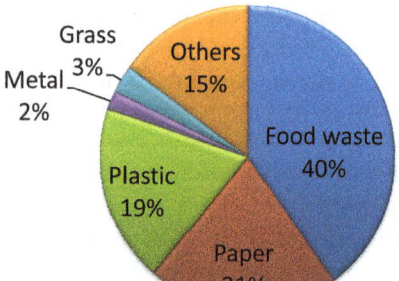

Figure 5.2 The composition of food waste in Hong Kong in 2011.[10]

waste (MSW), accounting for almost 40% of the MSW in 2011 (Figure 5.2).[10] In fact, most of the food waste in Hong Kong is from the domestic sector, contributing about 70% to the total food waste generation while the remaining 30% is from the C&I sector. However, it is also noticed that the C&I food waste has increased 280% from 373 tonnes in 2002 to 1050 tonnes in 2011. Therefore, it is anticipated that the food waste generation in Hong Kong will continue to rise, driven by the significant increase of the C&I food waste generation.

Following the re-organization of government bureaux on 1 July 2007, the Environment Bureau was formed, which is responsible for the formulation and implementation of environmental laws and policy. EPD is a branch of the Environment Bureau and has adopted a new structure based on three operational divisions, four policy divisions, a cross-boundary division, and a corporate affairs division. Figure 5.3 shows the divisions and groups related to waste management under the Environment Bureau. Although there is a division, namely the Waste Management Policy Division, designated for policy formulation, strategic planning and programme development in the field of waste management including policy for waste reduction and re-cycling, most currently discharged food waste is still dumped together with

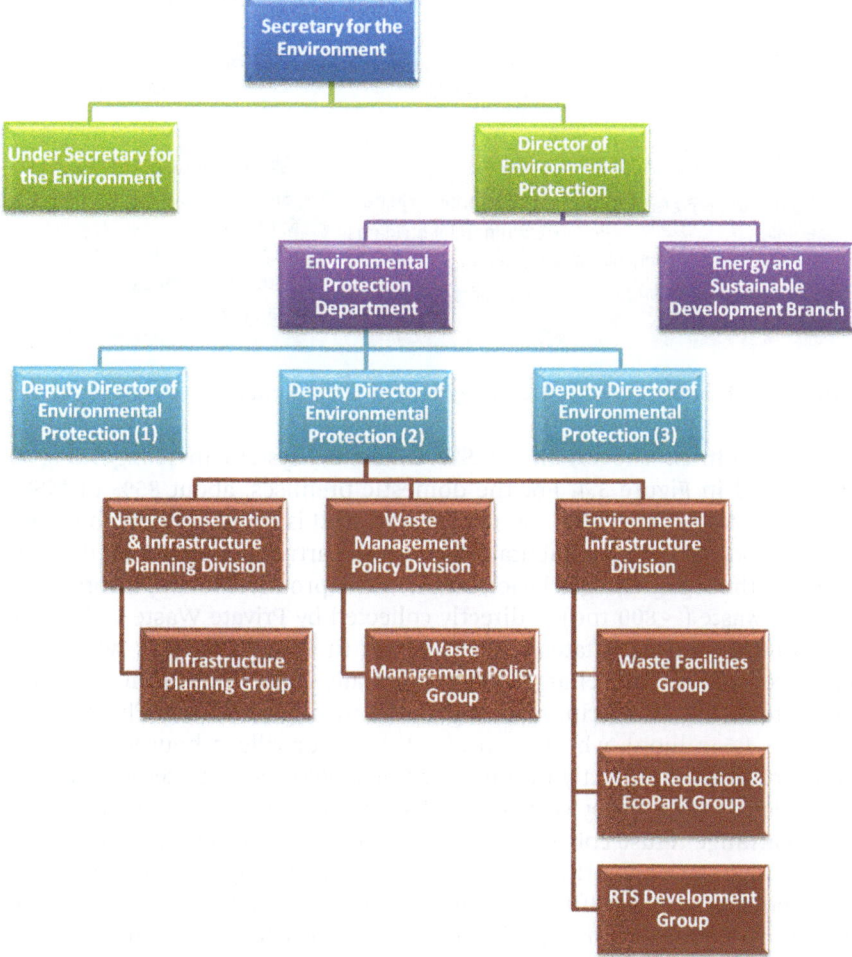

Figure 5.3 The divisions and groups related to waste management under the Environment Bureau.

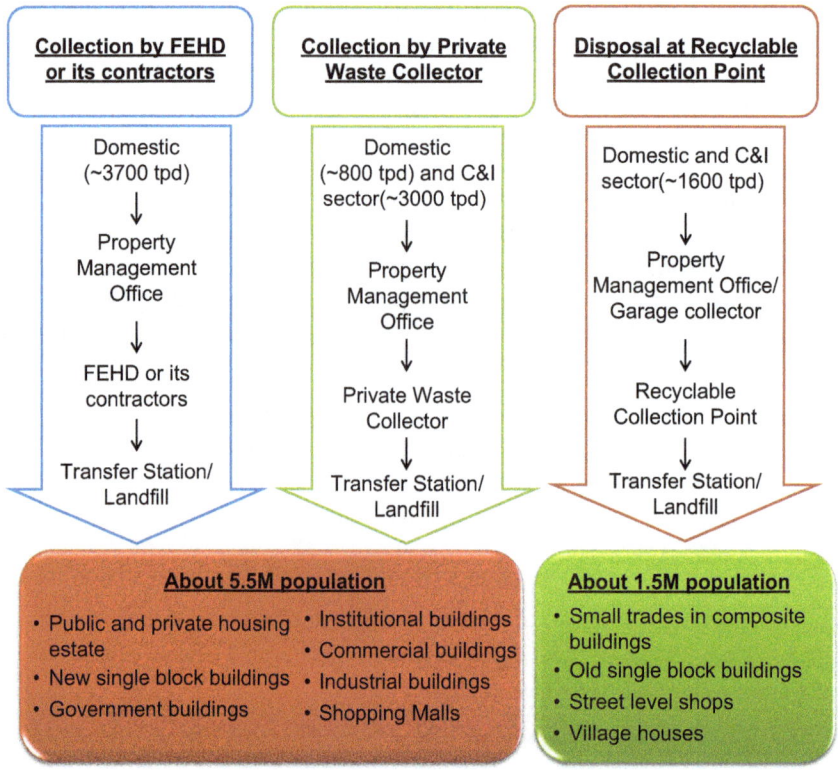

Figure 5.4   Existing MSW Collection System in Hong Kong.

MSW to landfills. The existing MSW collection system in Hong Kong is summarized in Figure 5.4. For the domestic premises, about 85% of MSW from domestic source (3700 tonnes per day, tpd) is transferred to the landfills by Food and Environmental Hygiene Department (FEHD) or its contractors without any charges levied on the waste producer. Then, a portion of domestic waste (~800 tpd) is directly collected by Private Waste Collectors from private housing estates, which do not fit in with FEHD's waste collection schedules, with charges. They normally collect both domestic and C&I waste in the same trip, resulting in the mixing of wastes. Third, some people (~1.5 million), who live in old districts or village houses scattered around rural areas, directly dump their household waste at Recyclable Collection Points (RCP) or by Garbage Collector since there is no management party to arrange refuse collection services. For C&I premises, as they are not serviced by FEHD, thus they normally engage cleansing contractors to deliver their waste to landfills or Refuse Transfer Stations (RTS) where the waste is compacted and containerized in purposely built containers for onward transportation to the strategic landfills.

The disposal of food waste in landfill was found to be the most economical and efficient option,[11] although it causes numerous problems including the generation of greenhouse gas (GHG), leachate and unpleasant odour. As food waste is buried and compacted under the ground, the decomposition of food waste under an anaerobic environment produces methane, which is a GHG found to be 21 times more powerful than carbon dioxide in terms of the ability of giving greenhouse effect. Also, since food waste contains a large portion of moisture, it significantly increases the leachate generation in landfill, which may damage the landfill liners and contaminate ground water. Thus, additional investments are made by the landfill contractors to treat and monitor the landfill gas and leachate before discharge. What is more, the decomposition creates a great deal of gases that causing unpleasant odours, such as H_2S and NH_3, which always draws complaints from the residents nearby.

5.3 Overview of Food Waste Valorization Technology

Many food waste valorization technologies with different approaches, including waste to energy, waste to soil, waste to biomass and waste to animal feed, are available worldwide. In Sections 5.3 and 5.4, some of these technologies are reviewed regarding their principles, advantages and disadvantages in terms of social, technological and environmental performance with a scope of applying these technologies in Hong Kong.

5.3.1 Waste to Soil: Composting

Generally, composting is a natural process of microbial decomposition of organic matter that occurs in aerobic conditions.[12] Food waste can be degraded into available nutrients for the growth of plants in the process, so composting therefore is a very common technology to return food waste to soil.

It is a dynamic process involving a rapid sequential activity by mixed microbial populations, which are mainly fungi and actinomycetes (fungi-like bacteria with long filaments spreading through the soil and decomposing the organic matter).[13] In the composting process, the three phases of mesophilic phase, thermophilic phase and the cooling and maturation phase, are followed. The microbial diversity varies in the different phases but the total number of microorganisms is consistent. First, the composting mass proceeds through the mesophilic phase for 3–4 days in a slightly acidic environment and ambient temperature at first. Easily degradable carbon sources like monosaccharides, lipids and starch are decomposed by mesophilic bacteria, which are dominated by thermophilic fungi. As the excess energy is released as heat and organic acids are formed during degradation, the temperature increases while the pH is decreasing during the mesophilic phase. Then, the more resistant compounds like natural polymers and proteins are degraded in the thermophilic phase by thermophilic actinomycetes. The degradation of proteins results in the liberation of ammonia and an increase in the pH. The high temperature kills pathogens in the thermophilic phase. Finally, the cooling and maturation phase carried out by both mesophilic bacteria and thermophilic bacteria lasts for between half to even several months in which harder materials, odours and toxic intermediates are decomposed. The whole process lasts for at least 4–6 months, in which different microorganisms transform food waste into CO_2, heat, water biomass and compost, which is a humus-like soil fertilizer.[14] The efficiency of composting can be optimized by various factors, including moisture content, temperature, pH, initial C/N ratio and aeration.[15] First, the moisture content in the composting mass can be adjusted to certain levels in accordance to the composting materials by the addition of water or bulking agent, which could absorb excess moisture and give structure to the composting mixture. High levels of moisture content can cause a lack of aeration and the leaching of nutrients while low levels of moisture decreases the microbial activity since microorganisms are only able to use organic molecules dissolved in water.[13] The initial C/N ratio is recommenced to be 25 to 30 since carbon is the energy source for the microorganisms while nitrogen is a component necessary for cell growth and functioning. Low levels of carbon and nitrogen can result in low degradation rate whereas high levels of carbon and nitrogen lead to excess heat release and the formation of ammonia gas respectively. Third, a temperature exceeding 60–65 °C would kill almost all microorganisms and end the decomposition while temperatures lower than the critical point would result in a low decomposition rate. Therefore, a temperature ranging up to 60 °C is recommended to achieve an optimal decomposition rate and to destroy pathogens, weed seeds and fly larvae. Fourth, as with temperature, the pH also controls the rate of microorganisms' activity. It should be slightly acidic at first and then become neutral or slightly alkaline. Finally, enough aeration should be provided, otherwise anaerobic digestion may take place instead of the aerobic composting process. All in all, the efficiency of composting can be optimized once a favourable environment is provided to the microorganisms. The general input flows and output flows of composting are summarized in Table 5.1. Food waste mixed with a bulking agent like sawdust undergoes the process which involves the usage of electricity due to the machinery equipment. 1000 kg of food waste can produce 250 kg of compost and 600 kg of wastewater, which is highly polluting as it consists of ammonia, bacteria and a high level of organic matter. Also, some screenings of solid debris are removed before or during the process.

5.3.2   Waste to Food Chain: Animal Feed Production

Feeding food waste directly to animals has long been practiced to reduce the cost of animal production. Some by-products or waste products, such as corn gluten meals, wheat middlings, used cooking oils and distiller grain, have been used as feedstuffs or protein sources throughout the centuries.[17] Numerous studies confirmed the feasibility of turning different kinds of food waste, such as meat waste, fruit and vegetables waste, restaurant waste and

Table 5.1 The main input flows and output flows of each option.

	Input			Output			
	Item	Value	Unit	Item	Value	Unit	Ref.
Composting	Food waste	1000^a	kg	Compost	250^a	kg	16
	Sawdust	179	kg	Wastewater	600	kg	
	Eleccricity	117.3	kWh	Screenings	59^a	kg	
Animal feed	Food waste	1000^a	kg	Dry feed	670	kg	27
	Electricity	204	kWh				
	Propane gas	12	m^2				
Incineration	Raw waste	1	g	Ash	0.22	g	29
	Natural gas	6.01E-05	g	Electricity	5.94E-04	kWh	
	Electricity	6.68E-05	kWh				
	Diesel	1.57E-04	g				
	$Ca(OH)_2$	3.20E-03	g				
	Activated carbon	2.50E-03	g				
	CaO	2.50E-02	g				
Anaerobic digestion	Food waste	1000^a	kg	Biogas	223	m^3	27
	Water	4.3	m^3	Electricity	281	kWh	
	Electricity	149	kWh	Compost	300^a	kg	

^{a}*Note*: The value is in wet basis.

Table 5.2   Feed and water intake, body weight change and feed efficiency of rats fed dry and fermented diets.[a,22]

Item	Diet			
	Dry	Aerobically fermented	Anaerobically fermented	Anaerobically fermented with 0.5% LAB culture added
Feed intake (g per day)	20.3	21.0	20.7	19.2
Water intake (mL per day)	31.8	30.9	30.5	31.5
Initial body weight (g)	197.1	197.0	195.5	196.8
Final body weight (g)	318.6	335.2	333.8	326.1
Average daily gain (g)	4.34	4.94	4.94	4.62
Feed efficiency (g/gain, g)	4.69	4.28	4.20	4.16

[a]Mean of 12 rats (4 cages with 3 rats per cage).

household waste, into animal feed.[2,18–20] Businesses utilizing food waste from the food manufacturing industry, like vegetable oils and soap stock in animal feed production, are already present in the world.

Usually, animal feed production involves separation, fermentation and dehydration. From the practice of a food waste recycler in Hong Kong, food waste is separated with reference to its nutritive values first.[21] Only food waste with high protein and carbohydrate content is used so as to produce animal feed with high nutrient contents, especially high protein content, with a high price than those with low nutritive values. Then, food waste could be processed by aerobic and anaerobic fermentation for better animal growth performance. According to a study, which was conducted to find out the differences in feeding dry or fermented (aerobically or anaerobically with or without lactic acid bacteria, *Lactobacillus salivarius*) restaurant food residue diets to animals, feeding aerobically or anaerobically fermented diets showed better animal performance as indicated by higher feed efficiency and rat growth rate (Table 5.2), which were attributed to the better protein conservation in diet during the fermentation process and to higher total tract digestibility of neutral detergent fibre and crude ash in the fermented food residue diets.[22] Finally, methods such as dehydration, pelleting, and extrusion were found to be feasible to produce animal feed.[23–26] From local experience, the whole process lasts for 1 to 7 days, which depends on the fermentation duration.[21] The general input flows and output flows of dry animal feed production are summarized in Table 5.1.[27] The conversion rate is 67%, so that 1000 kg of food waste can be converted into 670 kg of dry animal feed. Meanwhile, propane gas is used for the drying process and electricity is used for the operation of machinery like food waste crushing and mixing.

5.3.3   Waste to Energy: Incineration

Incineration is a well-developed but energy intensive technology involving combustion of waste materials by controlled burning to generate electricity.

The process generally is divided into incineration, energy recovery and air pollution control. First, the waste is fed into the furnace for combustion with a sufficient air supply, with residence time and a burning temperature of more than 850 °C to ensure complete combustion and to prevent the formation of dioxins and carbon monoxide. Then, for the energy recovery, the heat generated from the combustion of waste is used to produce steam in the boiler, which then drives the turbine to generate electricity. Lastly, an air pollution control system is installed to minimize the emission of air pollutants. A dry/wet scrubber neutralizes the acidic gases like SO_2 and HCl by spraying lime powder into the hot exhaust gas. A bag filter system removes the particulates and dust particles from the fuel gas. An activated carbon column adsorbs the heavy metals and organic pollutants such as PCB and VOC in the fuel gas.[28]

The general input flows and output flows of incineration are summarized in Table 5.1.[29] Food waste, together with the other solid waste, can be fed into the incinerator, with the addition of natural gas for initial start-up and maintenance of high combustion temperatures, hydrated lime and activated carbon for air pollution control. During the process, energy is recovered by producing electricity, and the volume of solid wastes is reduced up to 80% with the production of left over ash after burning.[29] Many countries, especially those with little landfill space, have chosen incineration to handle municipal solid waste, including food waste. From their practice, the ash after incineration is usually transported to landfills for disposal although it can be recycled form cement after solidification with cement agent.[28] As it is assumed that there should not be a large portion of organic matter due to the complete combustion, disposal of ash on landfill does not cause the formation of landfill gas. All in all, incineration is used globally as an efficient waste volume reduction measure to relieve the pressure of waste on landfill and expanding the life-span of landfill sites.

5.3.4 Waste to Energy: Anaerobic Digestion

Anaerobic digestion (AD) is a process involving the microbial decomposition of organic matter into methane, carbon dioxide, inorganic nutrients and compost in anaerobic environment. Four groups of microorganisms are involved: syntrophic bacteria, fermentative bacteria, acetogenic bacteria and methanogenic bacteria.[30] Methane is the ultimate product, which is regarded as a renewable energy source. Generally, the entire process takes 10–40 days, depending on the retention time of the fermentation, and can be divided into three stages: hydrolysis, acid forming and methanogenesis.[30] Firstly, hydrolysis is a reaction, catalysed by enzymes, in which the complex organic molecules are broken down into soluble monomers. In the case of hydrolysing organic material, the hydrolysis reaction is where organic waste is broken down into a simple sugar, in this case glucose, as can be represented in equation 5.1 in Table 5.3.[31] Then, the hydrolysed products are transformed into simple organic acids at the acid forming stage, and the

Table 5.3   Reactions involved in anaerobic digestion process.[31]

Reactions	Equations	
Hydrolysis	$C_6H_{10}O_4 + 2H_2O \rightarrow C_6H_{12}O_6 + 2H_2$	(5.1)
Acidogenesis	$C_6H_{12}O_6 \rightarrow 2CH_3CH_2OH + 2CO_2$	(5.2)
	$C_6H_{12}O_6 + 2H_2 \rightarrow 2CH_3CH_2COOH + 2H_2O$	(5.3)
Acetogenesis	$C_6H_{12}O_6 + 2H_2O \rightarrow 2CH_3COOH + 2CO2 + 4H_2$	(5.4)
	$CH_3CH_2OH + 2H_2O \rightarrow CH_3COO^- + 2H_2 + H^+$	(5.5)
	$2HCO_3^- + 4H_2 + H^+ \rightarrow CH_3COO^- + 4H_2O$	(5.6)
Methanogenesis	$2CH_3CH_2OH + CO_2 \rightarrow 2CH_3COOH + CH_4$	(5.7)
	$CH_3COOH \rightarrow CH_4 + CO_2$	(5.8)
	$CH_3OH + H_2 \rightarrow CH_4 + H_2O$	(5.9)
	$CO_2 + 4H_2 \rightarrow CH_4 + H_2O$	(5.10)
Overall	$C_6H_{10}O_4 + H_2O \rightarrow 3CO_2 + 3CH_4$	(5.11)

reactions are listed in Table 5.3. This stage of acid formation is facilitated by microorganisms known as acid formers. It comprises two reactions: acidogenesis and acetogenesis. During the acidogenesis, soluble organic products of the hydrolysis are transformed into simple organic compounds. Reactions of the organic waste involve the conversion of the glucose to ethanol (equation 5.2, Table 5.3) and conversion of the glucose to propionate (equation 5.3, Table 5.3). During the acetogenesis reactions, with the low concentration of H_2 and presence of hydrogen scavenging bacteria, organic acids can thus be formed. Reactions of the organic waste involve conversion of glucose to acetate (equation 5.4, Table 5.3), conversion of ethanol to acetate (equation 5.5, Table 5.3) and conversion of bicarbonate to acetate (equation 5.6, Table 5.3). Finally, methane is formed. Two-thirds of the methane production is from converting the acetic acid (equations 5.7 and 5.8, Table 5.3) or by the fermentation of alcohol (equation 5.9, Table 5.3) while the other one-third of methane production is from the reduction of the carbon dioxide by hydrogen (equation 5.10, Table 5.3).[31]

The general input flows and output flows of AD are summarized in Table 5.1.[27] Food waste, together with the other organic waste, can be fed into the fermentation tank for the production of biogas and compost. Water is added for the reactions and to sustain the growth of bacteria while electricity is used for the operation of machinery in crushing, mixing and dewatering. It was found that 1000 kg of food waste can generate approximately 300 kg of compost and 223 m^3 of biogas, which can be utilized to generate 281 kWh of electricity.[27]

5.3.5   Waste to Biomass Resource

In the last decade, the possibilities of using food waste as a sustainable resource for the production of bioenergy and biomaterials or as a chemical feedstock have drawn a lot of attention due to the foreseeable reduction in

Table 5.4 Chemical products extracted from food processing residues.[32]

Process	Chemical compound	Yield (%)	Ref.
Fish & crustaceans processing	Collagen	40–71	35
	Gelatin	18–39	36, 37
	Fish oil	96	38
Milk and Cheese production	Lactose	74	39
Wine, coffee, cocoa production	Tartaric acid	92.4	40
	Caffeine	0.77 (w/w)	41
	Polyphenols	78	42
Fruit & vegetable processing	Pectin	77.6	43
	Phenol	94.1–98.7	44
	Vitamin E (tocopherolquinone)	31–120	45

fossil fuel resources and its potential to reduce the carbon dioxide burden of the atmosphere. In fact, different kinds of food waste are generated throughout the life cycle. The food waste generated in the stage of food processing or production namely 'food processing residues' can be utilized as biomaterial and bioenergy.[32] For examples, pomace, mainly composed of cellulosic fibres, can be used as a structural material in the production of bricks or as an adsorbent in water treatment; waste cooking oil can be used to obtain biodiesel; cocoa bean shells can be used to produce some useful products, such as polyphenols, glue, pectins, a pigment and a flavour.[32] Waste citrus peels can be used for the combined production of D-limonene, pectin and flavonoids under low temperature hydrothermal microwave conditions.[33]

Also, food-processing residues are already used as a chemical feedstock by separation or extraction or fermentation. Some special compounds, which are always present in small amounts but are high in value, can be extracted from food processing residues. These compounds can be used for a huge number of special applications like detergents and cosmetic additives. Table 5.4 lists some major compounds that may be extracted from diverse food processing residues.[32] For the fermentation approach, most of the food-processing residues have a high nutritive value, which makes them useful for bacterial growth in fermentation. For example, bakery wastes were recently found to be useful for fermentative succinic acid production, with a yield of 0.55g per g of bakery waste.[34]

5.4 Advantages and Disadvantages of Food Waste Valorization Technologies

5.4.1 Waste to Soil: Composting

Generally speaking, composting is able to return food waste to the soil as a soil fertilizer, which can improve the yields and the crop quality by increasing the water holding capacity and water infiltration rates of the soil and also reducing the bulk density, erosion potential, usage of herbicides, pesticides and chemical fertilizer, and preventing GHG emissions from the production of the chemical fertilizer.[46] However, some limitations and disadvantages can be found. First, in view of the high population density in Hong Kong, the proper siting of composting facilities is quite difficult as it does not only require a relatively large area and also causes an odour nuisance. Then, input adjustment and long processing time restrict the efficiency and capacity of the composting facility. Also, agriculture in a cosmopolitan city such as Hong Kong is no longer a major industry, so making compost is not profitable because of the lack of market or demands. Environmentally, large amounts of highly polluted leachate can cause water pollution unless it is properly treated. Even so, it definitely increases costs for setting up wastewater treatment facilities or gives an added burden to existing sewage treatment plants. Last but not least, composting gives net greenhouse gas emission of 45.26 kg CO_2 per ton of waste and 5.06 kg CO_2 per ton of waste during the process of machine integrated composting and windrow composting respectively.[27]

5.4.2 Waste to Food Chain: Animal Feed Production

Socially, people have concerns on the safety of animal feed production, with worries about the presence of pathogens or parasites resulting in spreading of infectious diseases. Technically, it is possible to return food waste to the food chain so as to achieve a sustainable cycle of food waste generation and treatment. Also, the process is relatively simple and less equipment is involved than the other food waste valorization technologies and thus the processing time is short, about 1 to 7 days. However, from local experience, producing animal feed from food waste is relatively not a profit-making business because of the high additional costs resulting from the numerous hygienic measurement, high labour and land costs for food waste collection and separation of food waste with high nutritional value (Table 5.5).[21] For the environmental performance, the regenerated animal feed can substitute for commercial animal feed, and thus there is a net saving of the GHG emission of 125.58 kg CO_2 per ton food waste.[27]

Table 5.5 The nutrient composition of animal feed from HKOWRC.

Components	Quantity
Available carbohydrate (g per 100 g)	32
Energy (kcal per 100 g)	271
Total dietary fiber (g per 100 g)	32
Sodium (mg per 100 g)	140
Protein (g per 100 g)	21
Saturated fat (g per 100 g)	0.85
Sugars (g per 100 g)	4.8
Total fat (g per 100 g)	6.9
Water content (g per 100 g)	2.7
Ash (g per 100 g)	6.0

5.4.3 Waste to Energy by Incineration

Technologically, incineration is definitely a convenient, effective and efficient technology with reference to its short processing time, energy recovery, significant volume reduction and low requirement of the feeding materials. However, the cost of incineration is very high due to the high initial investment of an incinerator with burner and devices for pollution control, and also with high maintenance and operating costs. Public opposition is undoubtedly another major hurdle. According to a survey carried out in Tseung Kwan O in Hong Kong,[47] only 16% of those interviewed agreed with the use of incineration for MSW treatment, indicating that site selection of the incineration plant is a difficult task in some densely populated cities nowadays. What is more, if there is no appropriate air pollution control, incineration definitely deteriorates the air quality by producing large amount of air pollutants like CO_2, CO, SO_2, NO_x, N_2O, HCl, NH_3, HF, PM_{10} and dioxin (Table 5.6).[48] While NO_x, SO_2 and PM_{10} are common air pollutants contributing to the air pollution in many cities, food waste combustion definitely further worsens the air quality.

5.4.4 Waste to Energy by Anaerobic Digestion

An advantage of anaerobic digestion is that the AD facility usually does not attract public opposition. Since the process takes place in enclosed tanks, there will not be any odour emission to nearby residents. Also, the reasonable land requirement gives high flexibility of site selection. Environmentally, AD delivers great environmental performance in view of the outputs, which are biogas and compost. Biogas and compost are renewable energy and soil fertilizer respectively. Therefore, with the substitution of the traditional fuel and chemical fertilizers, net savings of GHG emission were found to be 48.25 kg CO_2 per ton food waste.[27] Moreover, AD is a sustainable technology to treat food waste as the rest of the waste can undergo anaerobic

Table 5.6   Air emission from incinerator with air pollution control.[48]

Pollutants	Emission (kg per tonne MSW)	Emission (kg per 2500 tonnes food waste)
CO_2	216	540 000
CO	0.02	50
SO_2	0.08	200
NO_x	0.7	1750
N_2O	0.31	775
HCl	0.04	100
NH_3	0.006	15
HF	0.0003	0.75
PM_{10}	0.004	10
Dioxins/furans	3.62×10^{-11}	9.05×10^{-8}

digestion to become renewable resources. There are no by-products or screenings being disposed to landfill in AD. Technologically, the processing time is reasonable and flexible, ranging from 10 to 40 days. Also, AD is capable to treat materials with high water content like sludge; therefore it is often adopted in sewage treatment plants to generate electricity for on-site use.

However, there are disadvantages. The yield of biogas production varies due to the composition of food waste and retention time. Highest methane yields are found to be with excess of lipids and longest retention time while there are inhibitory effects likely to occur with excess of lipid and proteins due to the volatile fatty acid accumulation and ammonium nitrogen.[49] Besides, there is extra attention and precaution for the leakage of CH_4, extra cost of maintenance and air pollution control measures are needed.

5.4.5 Waste to Biomass Resource

Generally speaking this process does not draw public opposition as the processes involved will not cause any nuisance or hazard to people. It can also improve the image of food and processing industry. Technologically, food processing residues are easily accessible and can be collected in large quantities. It effectively utilizes food waste as a rich carbon and nitrogen source for high-value products and also reduces dependence on fossil fuel resource and reduces the carbon footprint of the existing treatment. It also increases the sustainability of the food production chain.

Disadvantages include technological limitations like fluctuated supply of food waste, inherent heterogeneously variable composition, low yield of product and long running time for most cases constitute a challenge for the development of robust large scale, consistent industrial processes. Also, the high initial cost for acquisition of equipment and high maintenance fee contribute to a heavy economic burden to the operation of facilities. What is more, because considerable knowledge-based processing is involved, professionals and personnel from the technological sector are needed.

5.5 Food Waste Valorization Technologies in Hong Kong

Nowadays, apart from landfill, a small part of food waste is diverted to composting facilities and animal feed production plants established and operated by the Hong Kong government or private companies.[21,50] However, these facilities are usually on a pilot scale with limited capacity. In the future, the Hong Kong government has planned to adopt AD and incineration in two newly built waste treatment facilities respectively so as to conserve the landfill capacity.[50] However, these facilities could only help solving part of the food waste problems in Hong Kong.[50]

5.5.1 Composting

In fact, there are numerous composting facilities in different scales in Hong Kong. For the small scale facilities, composting machines with the capacity of 50 to 200 kg food waste per day are purchased and installed in estates subsidized by the Environment and Conservation Fund. For the large scale facilities, the Animal Waste Composting Plant in Ngau Tam Mei and the Kowloon Bay Pilot Composting Plant (KBPCP) funded by government convert animal waste and food waste into compost respectively. KBPCP is a pilot composting plant covering a 2 hectare area, started 2009, to recycle food waste by an in-vessel composting method.[50] The treatment capacity was up to 500 tonnes of food waste annually, generating 100–200 tonnes of compost, which was distributed to different parties like non-governmental organizations and private farms for free.

5.5.2 Animal Feed Production

In Hong Kong, there are a few private facilities converting food waste into animal feed. One of them is named the Hong Kong Organic Waste Recycling Centre (HKOWRC), which converts C&I food waste into swine feed. In HKOWRC, food waste collected by trucks undergoes three stages, which are separation, fermentation and dehydration, to become swine feed. First, food waste with high protein and carbohydrate content is selected in the separation stage by manpower. Then, the selected food waste is fermented with Effective Microorganisms (EM), which is a combination of 70 to 80 different types of beneficial microorganisms. The principal organisms in EM are usually photosynthetic bacteria, lactic acid bacteria, yeasts, actinomycetes and fermenting fungi. They assist one another for survival in a food chain system and thereby form a synergy that fights off pathogens and decomposer microorganisms and thereby EM is self-sterilizing.[51] The major advantage of EM for HKOWRC is to deodorize the food waste during the fermentation process. Finally, the fermented food waste is dehydrated at high temperatures of up to 600 °C (baking) for an hour to produce swine or fish feed. Usually, the whole process lasts for about 5 days to 7 days in which fermentation accounts for most of the time. According to the record given by HKOWRC, 15 tonnes of food waste are converted into 7 tonnes of animal feed daily.[23] The animal feed was found to contain 21% protein (see Table 5.5), which could meet the recommended protein content of swine feed at different growing stages.[52] The information of the HKOWRC is summarized in Table 5.7.

5.5.3 Incineration

From the 1970s to the 1990s, MSW used to be incinerated first and then dumped to landfill. According to the White Paper 'Pollution in

Table 5.7   Summary of the information of Hong Kong Organic Waste Recycling
Centre, Integrated Waste Management Facilities and Organic Waste
Treatment Facilities.[50]

	Hong Kong Organic Waste Recycling Centre	Integrated Waste Management Facilities	Organic Waste Treatment Facilities
Technology adopted	Animal feed production	Incineration	Anaerobic digestion
Capacity	15 tonnes per day	2500 tonnes per day	200 tonnes per day
Area	0.3 ha	10 ha	2 ha
Construction cost	HK$3 million	HK$15 million	HK$500 million
Animal feed generation	2555 tonnes per year	N/A	N/A
Expected electricity generation	N/A	480 million kWh per year	14 million kWh per year
Expected ash generation	N/A	720 tonnes per year	N/A
Expected compost generation	N/A	N/A	7000 tonnes per year
Expected net saving of GHG gas	N/A	440 000 tonnes CO_2 per year	50 000 tonnes CO_2 per year

Hong Kong – A Time to Act' issued by government, in light of the severe air
pollution resulting from the incineration, incineration was phased out in the
1990s.[53] However, the Hong Kong government recently announced the
building of an incineration facility namely Integrated Waste Management
Facilities (IWMF) at Shek Ku Chau in 2008. Information about the IWMF is
summarized in Table 5.7. The area of the plant is about 10 hectares while the
construction cost is about HKD 15 billion. It is expected that 2500 tonnes of
MSW can be treated everyday by the facility. Energy recovered from the MSW
could be turned to 480 million kWh of electricity annually, which is only
about 1% of the total electricity consumption in Hong Kong. Also, the daily
production of 720 tonnes of ash will be dumped to landfills. Although the
ash contains various hazardous substances, such as easily leachable heavy
metals, soluble salts and organic compounds, it is found to be safe after
being immobilized in cement.[54] However, the plans received a great deal of
concern about the deterioration of air quality and land reclamation by the
public, especially for the nearby residents. A comprehensive environmental
impact assessment on the construction and operation of the IWMF was
conducted and approved under the Environmental Impact Assessment
Ordinance in Hong Kong. Under this ordinance, the IWMF will comply with
the most stringent international emission standards, including particulates,
organic carbon, HCl, HF, SO_x, CO, NO_x, mercury, cadmium, total heavy metal
and dioxin.[55]

5.5.4 Anaerobic Digestion

The Hong Kong government announced the establishment of Organic Waste Treatment Facilities (OWTF) in which biological technologies (composting and anaerobic digestion) will be adopted to stabilize the organic waste and turn it into useful compost products and biogas for energy recovery. It is expected the first phase and second phase of the OWTF will be located at Siu Ho Wan of North Lantau and Shaling at North District respectively, together with treatment capacity of 400–500 tonnes of organic waste daily.[50] The phase 1 of OWTF will be in operation in 2015 with an estimated treatment capacity of 200 tonnes of organic waste daily, generating 20 tonnes of compost daily and 14 million kWh of electricity annually (see Table 5.7).[50]

5.6 Suitable Technologies for Hong Kong

In light of the difference between domestic food waste and C&I food waste, the most suitable technology for domestic and C&I food waste will be worked out with the consideration of the above findings. A more comprehensive plan for food waste management is needed in Hong Kong in view of the lack of landfills and increasing food waste productions.[1,10]

5.6.1 Commercial and Industrial Food Waste

In view of the significant amount of functionalized molecules in C&I food waste, decomposition by composting, AD and incineration are not suitable options since they do not take advantage of the valuable compounds in the food waste. Therefore, C&I food waste should be utilized to high-value products like high-protein animal feed, biomaterials or chemicals. General C&I food waste, which consists of large amount of nutrients like carbohydrates, protein, triglycerides and fatty acid, should be recycled to be high-protein animal feed while specific waste streams like bakery wastes, pomace and feathers are recommended to be utilized into the production of biomaterial or chemicals.

5.6.2 Domestic Food Waste

Since the composition of domestic food waste is variable and only contains a few valuable compounds, utilization of waste to biomass resource and animal feed production are obviously not capable to recycle domestic food waste.

5.6.2.1 Evaluation of Composting

Composting is the least suitable option for Hong Kong due to numerous reasons from social, technological and environmental aspects. First, composting facilities are likely to draw public opposition in Hong Kong since they require a relatively large area and often cause an odour nuisance. As Hong Kong is a densely populated city with very limited land, the siting of composting facilities is very difficult. Then, in view of the 2500 tonnes of domestic food waste generated every day in Hong Kong, composting is not a suitable option due to the low efficiency resulting from long processing time (120–180 days) and input adjustments (C/N ratio is 25–30 and the moisture content 50–55%). Also, the use of compost in Hong Kong is in low demand. If all the domestic food waste was converted to compost, assuming the conversion is 25% (see Table 5.1), then 500 tonnes of compost would be generated every day and would need to be exported outside of Hong Kong.

5.6.2.2 Evaluation of Incineration

Incineration is not suitable to treat domestic food waste in Hong Kong. Incineration is a convenient, effective and efficient technology with the shortest processing time and no input requirement, making incineration capable of treating the significant amount of domestic food waste generated daily. However, it is not a sustainable way in view of the problems of ash disposal and air pollutant emission. The use of incineration will deteriorate the air quality by producing large amount of air pollutants like CO_2, CO, SO_2, NO_x, N_2O, HCl, NH_3, HF, PM_{10} and dioxin.[48] Assuming 2500 tonnes of food waste is combusted, 1750 kg of NO_x, 200 kg of SO_2 and 10 kg of PM_{10} will be generated from the incinerator daily (see Table 5.6). While NO_x, SO_2 and PM_{10} are the air pollutants contributing most to air pollution in Hong Kong (with the annual emission of 114 000 tonnes for NO_x, 31 900 tonnes for SO_2 and 6220 tonnes for PM_{10} in 2011),[56] the combustion of food waste could increase the annual emission of NO_x, SO_2 and PM_{10} by 0.56%, 0.22% and 0.06%. It can be concluded that food waste combustion would definitely further worsen the air quality in Hong Kong. On the other hand, incineration is not accepted by the public in Hong Kong, which is a main reason of the phase out of incineration in 1990s.

5.6.2.3 Evaluation of Anaerobic Digestion

Comparatively, anaerobic digestion (AD) is found to be the best suitable option for Hong Kong in term of its environmental performance, output and convenience in food waste collection. AD is an environmentally friendly option, saving 48.45 kg CO_2 per ton waste GHG emission, and will not contribute to the air pollution as incineration does. For the output product, AD's output is mainly compost, biogas or electricity, which can be used locally or exported. In reference to the information of Organic Waste Treatment Facilities (OWTF) mentioned in Section 5.5.4, if all the domestic food waste (2534 tonnes per day) is sent to undergo AD, it is expected to produce electricity of approximately 182 million kWh annually which is enough to supply 39 000 households in Hong Kong. It definitely reduces the use of fossil fuel as well. Also, given the variable composition of domestic food waste in Hong Kong, AD provides much convenience for domestic food waste collection as it does not require food waste of specific composition, while composting and animal feed do. Therefore, AD is likely to be the most suitable option for Hong Kong.

5.7   Conclusions

In this review, the current food waste disposal system in Hong Kong and some food waste valorization technologies, including composting, animal feed, incineration, anaerobic digestion and waste to biomass resource, were introduced and analysed. To conclude, food waste recycling is definitely necessary to utilize such a valuable resource while waste to biomass and AD are recommended to treat C&I and domestic food waste respectively. There is no one single technology that can handle the food waste problem, so multiple technologies should be used. Waste collection or separation at source is deemed important, together with other aspects of waste management policy such as waste charging schemes and waste recycling subsidies from the government.

Acknowledgements

The authors gratefully acknowledge the generous contribution of the Hong Kong Organic Waste Recycling Centre in providing the associated information and data shown in this chapter.

References

1. EPD (Environmental Protection Department of HKSAR). West New Territories (WENT) Landfill. http://www.epd.gov.hk/epd/english/environmentinhk/waste/prob_solutions/msw_went.html (accessed on 10 April 2013).
2. M. L. Westendorf, *Food Waste to Animal Feed*, Iowa State University Press, Ames, Iowa, USA, 2000.
3. R. Zhang, H. M. El-Mashad, K. Hartman, F. Wang, G. Liu, C. Choate and P. Gamble, *Bioresour. Technol.*, 2006, **98**(4), 929–935.
4. W. Russ and R. Meyer-Pittroff, *Crit. Rev. Food Sci.*, 2004, **44**(1), 57–62.
5. C. S. K. Lin, L. A. Pfaltzgraff, L. Herrero-Davila, E. B. Mubofu, S. Abderrahim, J. H. Clark, A. A. Koutinas, N. Kopsahelis, K. Stamatelatou, F. Dickson, S. Thankappan, Z. Mohamed, R. Brocklesby and R. Luque, *Energy Environ. Sci.*, 2013, **6**, 426–464.
6. M. Grolleaud, Post-harvest Losses: Discovering the Full Story. Overview of the Phenomenon of Losses During the Post-harvest System. Rome, Italy, FAO, Agro Industries and Post-Harvest Management Service, 2002.
7. J. Parfitt, M. Barthel and S. Macnaughton, *Phil. Trans. R. Soc. B*, 2010, **365**(1554), 3065–3081.

8. EPA (United States Environmental Protection Agency), Terminology Services – Terms and Acronyms Report. http://iaspub.epa.gov/sor_internet/registry/termreg/searchandretrieve/termsandacronyms/search.do (accessed on 28 June 2013).

9. EPD (Environmental Protection Department), 2012. Public Consultation on Introduction of Charging to Reduce Waste. http://www.gov.hk/en/residents/government/publication/consultation/docs/2012/MSW.pdf.

10. EPD (Environmental Protection Department of HKSAR), 2011. Monitoring of Solid Waste in Hong Kong 2011, https://www.wastereduction.gov.hk/chi/materials/info/msw2011tc.pdf.

11. A. Tatsi and A. Zouboulis, *Adv. Environ. Res.*, 2002, **6**(3), 207–219.

12. Z. Milan, E. Sanchez, P. Weiland, R. Borja, A. Martin and K. Ilangovan, *Bioresour. Technol.*, 2002, **83**(3), 189–194.

13. C. G. Golueke, Principles of Composting, in *The Art and Science of Composting*, The JG Press Inc., Pennsylvania, USA, 1991, pp. 14–27.

14. M. Renkow and A. R. Rubin, *J. Environ. Manage.*, 1998, **53**(4), 339–347.

15. B. K. Adhikari, Urban Food Waste Composting, MSc Thesis, McGill University, Montreal, 2005.

16. M. H. Kim and J. W. Kim, *Sci. Total Environ.*, 2010, **408**(19), 3998–4006.

17. M. L. Westendorf, *Food Waste to Animal Feed*, Iowa State University Press, Ames, Iowa, USA, 2000.

18. R. O. Myer, J. H. Brendemuhl and D. D. Johnson, *J. Anim. Sci.*, 1999, **77**(3), 685–692.

19. N. P. Kjos, M. Øverland, E. Bryhni Arnkværn and O. Sørheim, *Acta Agric. Scand. Sect. A Anim. Sci.*, 2000, **50**(3), 193–204.

20. A. Garcia, M. Esteban, M. Marquez and P. Ramos, *Waste Manage.*, 2005, **25**(8), 780–787.

21. HKOWRC (Hong Kong Organic Waste Recycling Center), http://www.hkowrc.com/ (accessed on 5 April 2013).

22. Y. I. Kim, J. S. Bae, K. S. Jee, G. W. Lee, T. McCaskey and W. S. Kwak, *Asian-Australas. J. Anim. Sci.*, 2011, **24**(12), 1744–1751.

23. C. Tadtiyant, J. J. Lyons and J. M. Vandepopuliere, *Poult. Sci.*, 1989, **68**(Suppl. 1), 145.

24. R. O. Myer, T. A. DeBusk, J. H. Brendemuhl and M. E. Rivas, Initial Assessment of Dehydrated Edible Restaurant Waste (DERW) as a Potential Feedstuff for Swine, Res. Rep. Al-1994-2, Florida Agric. Exp. Sta., University of Florida, Gainesville, 1994.

25. M. E. Rivas, J. H. Brendemuhl, D. D. Johnson and R. O. Myer, Digestibility by Swine and Microbiological Assessment of Dehydrated Edible Restaurant Waste, Res. Rep. Al-1994-3, College of Agric., Florida Agric. Exp. Sta., Univ. of Florida, Gainesville, 1994.

26. J. Goldstein, *Biocycte J. Waste Recyc.*, 1995, **36**, 40–42.

27. M. Takata, K. Fukushima, N. Kino-Kimata, N. Nagao, C. Niwa and T. Toda, *Sci. Total Environ.*, 2012, **432**, 309–317.

28. H. K. Lam, W. M. Ip, J. P. Barford and G. Mckay, *Sustainability*, 2010, **2**, 1943–1968.

29. F. Cherubini, S. Bargigli and S. Ulgiati, *Energy*, 2009, **34**(12), 2116–2123.
30. K. Zieminski and M. Frac, *Afr. J. Biotechnol.*, 2012, **11**(18), 4127–4139.
31. K. Ostrem and N. J. Themelis, Greening Waste: Anaerobic Digestion for Treating the Organic Fraction of Municipal Solid Wastes, MSc Thesis, Columbia University, New York, 2004.
32. B. Mahro and M. Timm, *Eng. Life Sci.*, 2007, 7(5), 457–468.
33. L. A. Pfaltzgraff, M. Debruyn, E. C. Cooper, V. Budarin and J. H. Clark, *Green Chem.*, 2013, **15**(2), 307–314.
34. C. C. J. Leung, A. S. Y. Cheung, A. Y. Z. Zhang, K. F. Lam and C. S. K. Lin, *Biochem. Eng. J.*, 2012, **65**, 10–15.
35. N. Muralidharan, R. Jeya Shakila, D. Sukumar and G. Jeyasekaran, *J. Food Sci. Technol.*, 2013, **50**(6), 1106–1113.
36. M. S. Rahman, G. S. Al-Saidi and N. Guizani, *Food Chem.*, 2008, **108** 472–481.
37. B. Jamilah, K. W. Tan, M. R. Umi Hartina and A. Azizah, *Food Hydro-colloids*, 2011, **25**, 1256–1260.
38. N. Rubio-Rodríguez, S. M. de Diego, S. Beltrán, I. Jaime, M. T. Sanz and J. Rovira, *J. Supercrit. Fluids*, 2008, **47**, 215–226.
39. R. R. de Souza, R. Bergamasco, S. C. da Costa, X. Feng, S. H. B. Faria and M. L. Gimenes, *Chem. Eng. Process.*, 2010, **49**, 1137–1143.
40. B. Rivas, A. Torrado, B. Moldes and J. M. Domínguez, *J. Agric. Food. Chem.*, 2006, **54**, 7904–7911.
41. A. Shalmashi, M. Abedi, F. Golmohammad and M. H. Eikani, *J. Food. Process. Eng.*, 2010, **33**, 701–711.
42. A. El-Abbassi, A. Hafidi, M. C. Garcia-Payo and M. Khayet, *Desalination*, 2009, **245**, 670–674.
43. M. Pourbafrani, G. Forgács, I. S. Horváth, C. Niklasson and M. J. Taherzadeh, *Bioresour. Technol.*, 2010, **101**, 4246–4250.
44. T. Jerman, P. Trebše and B. Mozetič Vodopivec, *Food Chem.*, 2010, **123**, 175–182.
45. B. Isso and D. Ryan, *Eur. J. Lipid. Sci. Tech.*, 2012, **114**, 927–932.
46. S. Brown, M. Cotton, S. Messner, F. Berry and D. Norem, 2009. Methane Avoidance from Composting, http://faculty.washington.edu/slb/docs/CCAR_Composting_issue_paper.pdf.
47. K. M. Chan and K. W. Fan, A Survey of Opinions from Residents in Tseung Kwan O on the Nuisances of the Southeast New Territories (SENT) Landfill, International Conference on the Siting of Locally Unwanted Facilities: Challenges and Issues, pp. 125–132.
48. H. K. Jeswani, R. W. Smith and A. Azapagic, *Int. J. LCA*, 2012, **18**(1), 218–229.
49. L. Neves, E. Goncalo, R. Oliveira and M. M. Alves, *Waste Manage.*, 2007, **28**(6), 965–972.
50. EPD (Environmental Protection Department of HKSAR), Development of Food Waste Treatment Facilities in Hong Kong http://www.epd.gov.hk/epd/english/environmentinhk/waste/prob_solutions/WFdev_OWTF.html (accessed on 14 April 2013).

51. W. Esatu, A. Melesse and T. Dessie, *Afr. J. Agric. Res.*, 2011, **6**(16), 3841–3846.

52. Clemson University, 1995. Swine Feeding Suggestion. http://www.clemson.edu/psapublishing/pages/ADVS/EC509.pdf.

53. EPD (Environmental Protection Department of HKSAR), White Paper, Pollution in Hong Kong – A Time to Act. http://www.epd.gov.hk/epd/english/resources_pub/policy/files/White_Paper-A_time_to_act.pdf (accessed on 10 April 2013).

54. H. S. Shi and L. L. Kan, *J. Hazard. Mater.*, 2009, **164**, 750–754.

55. EPD (Environmental Protection Department of HKSAR), Emission Standards – Overseas and Hong Kong. http://www.epd.gov.hk/epd/english/environmentinhk/waste/prob_solutions/WFdev_emission.html.

56. EPD (Environmental Protection Department of HKSAR). Hong Kong Air Pollutant Emission Inventory. http://www.epd.gov.hk/epd/english/environmentinhk/air/data/emission_inve.html (accessed on 14 April 2013).

CHAPTER 6

Advanced Generation of Bioenergy

OLUWAKEMI A. T. MAFE, NATTHA PENSUPA,
EMILY MAY ROBERTS AND CHENYU DU*

School of Biosciences, Sutton Bonington Campus, University of
Nottingham, Loughborough, LE12 5RD UK
*Email: Chenyu.du@nottingham.ac.uk

6.1 Biofuel Introduction

Due to population growth and industrial expansion, the world's energy
consumption has increased dramatically over the past few decades. Currently, over 80% of the world's energy is derived from fossil fuels – crude oil,
natural gas and coal. According to the BP 2030 Energy Outlook projection,
global population is expected to rise to around 8.5 billion by 2030, resulting
in a 40% increase in the energy demand by 2030.[1] It is generally accepted
that fossil fuels are finite and cannot supply energy at its current rate in the
foreseeable future. Alternative renewable energy must therefore be considered in order to meet the increasing demand on energy.

Similar to wind and solar energy, biofuel has attracted increasing attention worldwide as a desirable type of renewable energy. Biofuel has several
distinctive advantages over other types of sustainable energy, some of which
are: biofuel can be transported using the existing distribution network, it can
be pumped into most existing vehicles with minimum modifications and
most importantly it helps to reduce CO_2 emission.

Biofuel is defined as a solid, liquid or gaseous fuel that is derived from
relatively recent dead biological materials. The 'relatively recent dead

RSC Green Chemistry No. 27
Renewable Resources for Biorefineries
Edited by Carol Sze Ki Lin and Rafael Luque
© The Royal Society of Chemistry 2014
Published by the Royal Society of Chemistry, www.rsc.org

biological materials' is to distinguish biofuels from fossil fuels which are derived from biological materials that died millions of years ago. In the year 2000, the amount of biofuel produced worldwide was around 16 billion litres and this figure increased to more than 100 billion litres in 2010.[2] The two main types of biofuels are bioethanol and biodiesel. Bioethanol is produced from sugary crops, starchy crops or lignocellulosic matter *via* yeast or bacterial fermentation. Biodiesel is generated from pure or waste plant, animal or algae oil *via* esterification or transesterification. Besides bioethanol and biodiesel, biofuels also include biobutanol, biogas, biohydrogen, pyrolysis bio-oil and other types of biofuels. Bioethanol is however the most widely produced biofuel and is often blended with gasoline (petrol) in volumes up to 5% (E5) in Europe, 10% (E10) in the United States and 25% (E25) and even up to 100% (E100) in Brazil.

6.2   First Generation Bioethanol Production

First generation bioethanol production process uses cereal grains or sugar crops as the starting material for bioethanol production. Globally, corn and sugarcane are the main biomass crops used as the United States and Brazil currently dominate the world's production of bioethanol using corn and sugarcane, respectively. In Europe, wheat is the dominant biomass used for the first generation bioethanol production. The technology of the first generation bioethanol production process is well-established in the fuel sector.

Saccharomyces cerevisiae is the dominant microorganism in the first generation of fuel ethanol production. In recent years, the worldwide bioethanol production reached around 80 billion liters per year.[3] In a typical industrial scale bioethanol fermentation process using *Saccharomyces cerevisiae*, around 8–14% (v/v) ethanol is produced and the glucose to bioethanol yield is usually over 90% of the theoretical yield. In some processes, simultaneous saccharification and fermentation is applied, in which α-amylase/glucoamylase is mixed with *Saccharomyces cerevisiae* and starchy raw materials. Most of yeast cells harvested in the fermentation are recycled and sent back in order to enhance the cell concentration in the fermenter. Around 5–10% yeast cells end up in Dried Distillers Grains with Solubles (DDGS), which could be sold as animal feed.

The fermentation of starchy or sugar crops to bioethanol is performed using a series of different processes which are dependent on the raw material used. A general bioethanol process includes milling, liquefaction, saccharification, fermentation, distillation and dehydration, as shown in Figure 6.1.

6.2.1   Milling

The milling process is applicable to both the sugar and starch crops. In the case of sugar crops, the sugar juices are extracted from the bagasse, which is

Starch or sugar crops

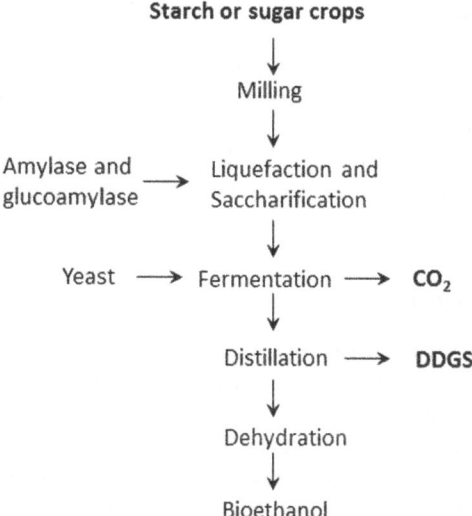

Figure 6.1 Schematic diagram of a typical first generation bioethanol production process.

then sent to the fermentation process while the bagasse is burnt in boilers for the generation of steam and electrical energy.[4] For starchy crops, the cereal grains are firstly milled and then the starch is extracted *via* either dry or wet milling processes.[5]

6.2.2 Liquefaction and Saccharification

The liquefaction and saccharification steps are required for the starchy crops. In these two processes, α-amylase and glucoamylase are added respectively to convert starch into glucose. These two processes are also collectively known as hydrolysis. It should however be noted that some bioethanol plants use acid instead of enzymes for the hydrolysis process.

6.2.3 Fermentation

The fermentation of the sugars to bioethanol is generally performed anaerobically using commercial yeasts such as *Saccharomyces cerevisiae*. Heat, CO_2 and a beer solution (fermentation broth) containing around 8–14% ethanol are the products of the fermentation process. The general equation for the fermentation reaction is shown below. The theoretical yield of bioethanol from this reaction is 0.51g per g of the consumed sugars while the actual yield obtained is around 90–95% of the theoretical yield.

$$C_6H_{12}O_6 \xrightarrow{\text{Microorganism}} 2C_2H_5OH + 2CO_2 + \text{heat}$$

6.2.4 Distillation

The beer solution obtained from the fermentation process then undergoes distillation in which the bioethanol is separated from other materials contained in the solution, thereby concentrating the bioethanol. Ethanol and water form an azeotrope at 95.57% ethanol (wt.) with a minimum boiling point of 78.2 °C, implying that more than 95.57% of ethanol concentration cannot be achieved by simple distillation.

6.2.5 Dehydration

A maximum concentration of only 95.57% can be achieved *via* distillation while a concentration of 99.5% is required for bioethanol to be blended with petrol. Because of this, after the distillation process, the bioethanol then moves on to the dehydration process to be further concentrated while the unfermented solids are further processed to DDGS or gluten feed/gluten meal. In the dehydration process, the $\sim 5\%$ (wt.) water contained in the ethanol is further reduced to less than 1% (wt.) with molecular sieves (*e.g.* zeolite). This process is important as when it comes to blending the bioethanol with gasoline, the high water content can result in a phase separation thereby resulting in engine malfunction.

6.3 Introduction of Advanced Generation Bioethanol Production

Currently, bioethanol is predominately synthesized *via* the first generation production processes, where food-based crops are used as the starting materials. However, converting food materials to biofuel triggered concerns about global food security and that significantly affected the public acceptance of biofuel. Research into biofuel has therefore been focusing on the development of advanced generation biofuel production processes using inedible materials. Within these non-food biofuel processes, production of bioethanol from lignocellulosic raw materials is one of the most promising options. Lignocellulosic raw materials are abundant, including wheat straw, corn stover, sugarcane bagasse, sawmill and paper mill discards, dedicated energy crops and municipal lignocellulosic and paper waste. These waste materials and dedicated energy crops allay the first generation food versus fuel fears as they are inedible, and are less expensive than the sugary and oleaginous plants. They are available in large quantities and do not require the use of additional land. Other benefits of the advanced generation biofuel production include: the development of rural areas in developing nations, the use of waste materials and abandoned land and the higher GHG reduction value compared to first generation biofuel production. Bioethanol produced from lignocellulosic biomass is believed to produce 75% less CO_2 than gasoline whereas that produced from sugarcane or maize reduces the CO_2 levels by 60%.[6]

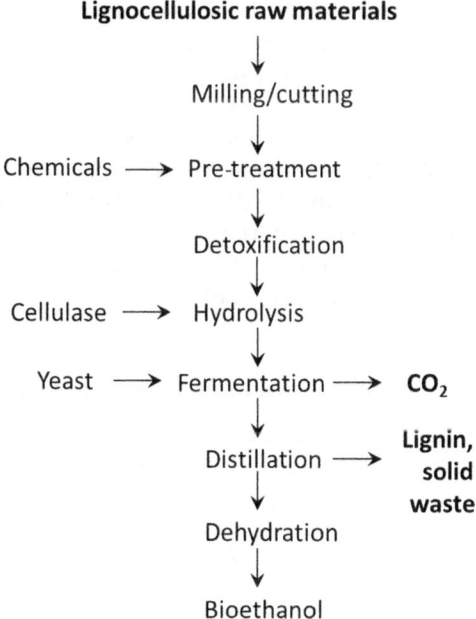

Lignocellulosic raw materials

↓

Milling/cutting

↓

Chemicals ⟶ Pre-treatment

↓

Detoxification

↓

Cellulase ⟶ Hydrolysis

↓

Yeast ⟶ Fermentation ⟶ **CO₂**

↓

Distillation ⟶ **Lignin, solid waste**

↓

Dehydration

↓

Bioethanol

Figure 6.2 A schematic diagram of a potential advanced generation bioethanol production process.

6.3.1 Advanced Generation Bioethanol Production Process

In principle, the lignocellulose to bioethanol production process is similar to the first generation bioethanol process. However, unlike the production of bioethanol from sugary or starchy materials, advanced generation bioethanol production requires an additional pre-treatment step to overcome the recalcitrance of the lignocellulosic raw materials and increase the surface area.[7] The biomass is first dried and cut before being milled to reduce the particle size. After which it undergoes pre-treatment for the hydrolysis process (usually enzymatic). The enzymes used are generally cellulases and other glycosyl hydrolases (GHs), as opposed to the amylases and glucoamylases used in first generation. The hydrolysate obtained which contains mainly simple sugars is then fermented to bioethanol. Lastly, the bioethanol is distilled and dehydrated just as in the first generation bioethanol production process to obtain the required specification. Figure 6.2 shows a typical advanced generation bioethanol production process.

6.4 Lignocellulosic Feedstock

The conversion of lignocellulosic biomass to bioethanol is, however, a challenging one as the raw material consists of complex polymers, namely cellulose, hemicellulose and lignin. These complex polymers are structured in such a way that causes the biomass material to be resistant to biological, chemical and physical attack. Moreover, the cellulose, hemicellulose and lignin compositions vary from biomass to biomass depending on the species, locations, seasons and harvest methods (See Table 6.1). Typical proportions (dry weight) of cellulose, hemicellulose and lignin in lignocellulosic biomass are 30–50%, 20–40% and 10–25%, respectively. As the composition and structure of lignocellulosic materials varies, the technology used in extracting energy from such materials is therefore more sophisticated than the first generation biofuels production process.

Table 6.1 Cellulose, hemicellulose and lignin contents in common lignocellulosic materials.

Lignocellulosic materials	Cellulose (%)	Hemicellulose (%)	Lignin (%)	Ref.
Barley straw	39.2	21.5	22.9	11
	33.8	21.9	13.8	12
Cotton stalks	58.5	14.4	21.5	12
Corn cobs	45	35	15	13
	33.7	31.9	6.1	12
Corn stalks	35.0	16.8	7.0	12
Corn stover	37.12	24.18	18.20	14
Eucalyptus saligna	48.07	12.69	26.91	14
Grasses	25–40	25–50	10–30	13
Hardwood stems	40–75	10–40	15–25	13
	40–55	24–40	18–25	15
Leaves	15–20	80–85	0	13
Monterey pine	41.70	20.50	25.90	14
Newspaper	40–55	25–40	18–30	13
Oat straw	32.9	21.7	18.1	11
	39.4	27.1	17.5	12
Paper	85–99	0	0–15	13
Rice straw	32–47	19–27	5–24	13
	36.2	19.0	9.9	12
Rye straw	25.9	21.5	27.1	11
	37.6	30.5	19.0	12
Softwood stems	30–50	25–40	25–35	13
	45–50	25–35	25–35	15
Soya stalks	34.5	24.8	19.8	12
Sugarcane bagasse	40	24	25	13
	40.0	27.0	10.0	12
Sunflower stalks	42.1	29.7	13.4	12
Switchgrass	33.75	27.04	16.80	14
	45	31.4	12.0	15
Waste paper from chemical pulps	60–70	10–20	5–10	13
Wheat straw	30	50	15	13
	32.4	19.1	21.3	11
	32.9	24.0	8.9	12

6.4.1 Cellulose

Cellulose consists of a linear chain of D-glucose monomers which are linked by β-1,4-glycosidic bonds. In plant cells, cellulose is present in both crystalline and amorphous forms. A crystalline and strong matrix structure is formed from the hydrogen linkages between cellulose layers. These tightly packed structures are called microfibrils and are extremely robust and resistant to degradation.[8]

6.4.2 Hemicellulose

Hemicellulose is a highly branched and complex polymer which contains pentoses (xylose and arabinose), hexoses (glucose, galactose, mannose, rhamnose, fucose), and uronic acids (glucuronic acid and galacturonic acid).[9] Xylan is the most common component of hemicelluloses. It contains mainly xylose (nearly 90%) that branches out to arabinose, galactose and other sugar compounds.[10] Depending on the type of feedstock, branch frequencies vary.[8] Hemicellulose coats the cellulose fibrils and forms bridges that hold the cellulose fibrils in place preventing them from collapsing. The branches in hemicellulose structures prevents them from forming microfibrils.[10]

6.4.3 Lignin

Lignin is a very complex molecule that consists of different types of phenolic monomers such as *p*-coumaryl, coniferyl and synapyl alcohols. It fills the spaces in the cell wall between the cellulose and the hemicelluloses matrix and is tightly bound to them *via* hydrogen and covalent bonds. This improves the plant's rigidity and compactness as well as its resistance against microbial attack (by restricting enzymatic access to the cellulose and hemicellulose). Lignin is amorphous with no fixed structure and therefore can only be represented by a hypothetical formula.

6.5 Pre-treatment

The pre-treatment step is crucial in overcoming the recalcitrance of lignocellulosic biomasses for biofuel production. For that reason, there are certain requirements that an effective and economical pre-treatment process should meet, such as, avoiding destruction of the hemicellulose and cellulose as well as formation of inhibitors, reducing the cost of the material for the reactor and consuming as little as possible chemicals (preferably a cheap chemical).[16] During the pre-treatment process, the carbohydrates are extracted and made more accessible to the following enzymatic hydrolysis and fermentation steps. The pre-treatment stage not only grants access to the sugars trapped inside of the cross-linking structure but also improves the rate of production by reducing the degree of cellulose polymerization thereby enhancing the total bioethanol yield. As there is no typical cell wall and hence no typical biomass structure, there is no typical pre-treatment process. There are several pre-treatment processes that have been developed for the biofuel formation process and the effectiveness of these individual pre-treatments is dependent on both the operating conditions and biomass composition. The pre-treatment processes can be classified into the following categories; physical, chemical and thermochemical, and biological pre-treatments.

6.5.1 Physical Pre-treatment

Physical pre-treatments are essentially processes that reduce the size or disrupt the structure of the biomass substrate in order to increase the surface area. For instance, the milling process is one such example which is also used in the production of bioethanol from sugarcane and corn.

6.5.1.1 Mechanical Pre-treatment

Mechanical pre-treatment or comminution is a pre-treatment process which reduces the size of material by chipping, grinding and milling. The aim of this is to reduce the particle size and increase the accessible surface area for enzymes. Milling pre-treatment of sugarcane bagasse for enzymatic hydrolysis showed that glucose and xylose hydrolysis yields from pre-treated bagasse were 70% higher than that of untreated bagasse.[17] Similar research was carried out to investigate the effect of particle size on hemicellulose conversion.[18] The result showed that the corn stover particles with sizes of 100 to 200 mesh (0.149 to 0.125 mm) gave higher hemicellulose conversion yield than the un-treated corn stover. It is generally accepted that small particle size benefits the hydrolysis process.[19] However, fine powder may not be desirable in commercial scale bioethanol production process due to higher energy consumption, capital cost and operating cost associated with the particle size reduction.

6.5.1.2 Freezing Pre-treatment

Freezing pre-treatment uses ice particles generated at extremely low temperatures to break or damage the plant cell wall. Chang *et al.* studied freeze pre-treatment on rice straw at $-20\ °C$ under atmospheric pressure to enhance enzymatic conversion. The results showed that freeze pre-treatment increased enzymatic conversion yield from 48% for untreated straw to 84% for freeze-treated rice straw.[20]

6.5.1.3 Irradiation Pre-treatment

Ionizing irradiation is another type of physical pre-treatment process that disrupts the structure of lignocellulosic biomass to form free fermentable sugars. The efficiency of irradiation pre-treatment depends on the lignocellulose structure and radiation dose. The effect of gamma irradiation pre-treatment on wheat straw had been studied by Yanga *et al.*[21] The results showed that gamma irradiation could break down the structure of wheat straw. It was also observed that an increase in irradiation doses resulted in an increase in weight loss and size reduction. Combined irradiation with alkali pre-treatment or dilute acid hydrolysis were also investigated.[22-24] Both results were similar.

6.5.2 Chemical and Thermochemical Pre-treatment

6.5.2.1 Acid Pre-treatment

Acid pre-treatment involves the use of either concentrated or dilute acids to break the lignin–hemicellulose bonds in the lignocellulosic materials. Acids, such as sulfuric acid and hydrochloric acid, split the matrix structure of the lignocellulosic material into cellulose, hemicellulose and lignin before further reducing hemicellulose and cellulose into simple sugars.

Dilute acid is the most widely investigated acid pre-treatment method and it is generally considered to be the most commercial feasible pre-treatment process. Sulfuric acid is the most commonly studied acid. The acid concentration is usually kept in the range of 0.2–5% and the temperature used is between 121 and 220 °C, with a reaction time of 10–30 minutes.[25,26] In the dilute acid pre-treatment process, after complete hemicellulose removal has been achieved, the acid continues to hydrolyse the hemicellulose to simple sugars which is favourable. However, the acid may further degrade the sugars to generate strong fermentation inhibitors, such as furfural (produced from the degradation of pentose sugars) and hydroxymethyl furfural (HMF, produced from the degradation of hexose sugars). Various biomass feedstocks have been degraded using dilute acid pre-treatment, *e.g.* hard wood (aspen), soft wood (balsam), rice silver grass, rice straw, corncobs and sugarcane bagasse.[27-30]

Concentrated acids such as concentrated phosphoric acid, sulfuric acid and hydrochloric acid are also used for pre-treating lignocellulosic biomass. The advantages of the concentrated acid pre-treatment process are relative high sugar recovery yield and low operation temperature, *e.g.* 50 °C. On the downside, the concentration of the acid used is very high, ranging from 30 to 70%. This high concentration together with the high temperature makes the acid solution extremely corrosive, leading to high capital costs in equipment. The concentrated acid pre-treatment is also relatively longer than the dilute acid pre-treatment; in addition, the acid needs to be recovered and recycled in order to make the process economically feasible. Zhang *et al.* reported the impact of concentrated acid pre-treatment on microcrystalline cellulose using 85% phosphoric acid at 50 °C for 10 hours.[31] After the pre-treatment, the microcrystalline cellulose was in the amorphous form and the sugar yield from the hydrolysis was over 3-fold that obtained using a non pre-treated substrate.[31]

6.5.2.2 Alkaline Pre-treatment

Alkali pre-treatment is effective in removing lignin by degrading the ester and glycosidic side chains in the lignin structure. Moreover, at high alkali concentrations (5–20%, wt), cellulose swelling, partial de-crystallization of cellulose and partial solvation of hemicellulose occur.[32] The most common alkaline pre-treatment agents are sodium hydroxide (NaOH), calcium hydroxide (Ca(OH)$_2$), potassium hydroxide (KOH) and ammonium hydroxide (NH$_4$OH). Alkali pre-treatment can be carried out at ambient temperature, but it takes several hours to days, whereas acid pre-treatment only takes a few minutes to hours.[32,33] In comparison with acid pre-treatment, alkali pre-treatment causes less sugar degradation, thus produces less furfural and HMF. Out of all the alkali agents, NaOH has been the most extensively studied while Ca(OH)$_2$ is considered to be the least expensive hydroxide per kilogram. An added advantage of Ca(OH)$_2$ pre-treatment is that the salt produced, CaCO$_3$, can be recovered by precipitation.[32] Alkali agents have been used to pre-treat various lignocellulosic materials such as oil palm empty fruit brunches, barley hull, sorghum straw and rice hull.[34–37]

Another type of alkaline hydrolysis is 'soaking in aqueous ammonia' (SAA), in which aqueous ammonia reacts with biomass at low temperatures. The aim of SAA is to remove the lignin in the raw material by minimizing its interaction with hemicellulose, thus providing more surface area for the enzymes.[15] Nguyen *et al.* explored the pre-treatment of rice straw with ammonia for the conversion of lignocellulose to fermentable sugars.[38] The result showed that a SAA process using a 10% (v/v) ammonia solution, at 100 °C reacting for 6 hours, resulted in a higher cellulose recovery yield of 57.4% and a lower lignin content of less than 10% in the hydrolysate than those of untreated rice straw (36.8% cellulose recovery yield and 15.8% lignin content).

6.5.2.3 Steam Explosion Pre-treatment

Steam explosion or steam pre-treatment is a pre-treatment process in which the lignocellulosic biomass is treated with high pressure and saturated steam followed by a dramatic decrease in the pressure at the end of the pre-treatment. During the process, the acetic acid and the other organic acids formed from the acetyl or other functional groups in the biomass, hydrolyses the hemicellulose while the lignin is condensed into small droplets due to the high temperatures of the reactor.[32] The sudden change in pressure makes the material explode, causing the glycosidic bonds in hemicellulose to break thereby removing the lignin from the matrix and increasing the cellulose surface area.[39] The factors that influence the steam explosion pre-treatment include pressure, retention time, temperature, biomass particle size and moisture content. The temperature and pressure in a common steam explosion pre-treatment are 160–260 °C and 0.69–4.83 MPa respectively, with a short retention time from a few seconds to minutes. Sulfur dioxide (SO_2), carbon dioxide (CO_2) or acids such as sulfuric acid (H_2SO_4) can also be added to the steam explosion process to improve the hydrolysis of hemicelluloses. This option is however necessary for softwoods as the steam explosion process alone is not as effective as it used for hardwoods and agricultural residues. SO_2-impregnated steam explosion is specifically effective in digesting softwoods.[40]

6.5.2.4 Ammonia Fibre Explosion

In ammonia fibre explosion (AFEX) pre-treatment, the biomass is subjected to liquid anhydrous ammonia at high pressures (1.7–2.1 MPa) and moderate temperatures (40–140 °C) for a period of time after which the pressure is suddenly released, just like in the case of steam explosion.[41] Due to the moderate temperatures used, the energy input is less than that required for other pre-treatment methods that employ high temperatures and therefore the cost associated with this process is relatively lower. However, the cost of ammonia itself, its recovery and neutralization treatment add to the overall cost.[15] Zheng *et al.* reported an optimum condition for a typical AFEX process: 1 to 2 kg of ammonia per kg of dry biomass at 90 °C for 30 minutes.[40] During the AFEX pre-treatment, hydrolysis of the hemicellulose, ammonolysis of the glucuronic cross-linked bonds and partial decrystallization of the cellulose occur. However, only a small amount of the hemicellulose is solubilized due to the degradation and deacetylation of the hemicelluloses.[32,33] As a result, the hemicellulose and lignin are not removed. Advantages of the AFEX process include a lower formation of degradation products, a complete recovery of the solid material and a reduction in lignin impact on enzymatic hydrolysis.[33] The AFEX technology has successfully been applied to pre-treat herbaceous and agricultural residues like coastal Bermuda grass, corn stover, bagasse, forage sorghum, sweet sorghum bagasse, rice straw and switch grass.[42–47] On the other hand, the AFEX technology is not very effective with lignocellulosic materials containing high lignin content, such as hardwoods, softwoods, newspaper and aspen chips.[32,40]

6.5.2.5 Carbon Dioxide Explosion Pre-treatment

Carbon dioxide explosion is a pre-treatment process that uses supercritical carbon dioxide to break down the biomass structure. In aqueous solution, carbon dioxide forms carbonic acid which depolymerizes lignocellulosic materials. As a small molecule, carbon dioxide can penetrate into the pores of the biomass better than ammonia. When carbon dioxide explodes due to the change of pressure, it breaks the cellulosic structure. This process is usually operated under high pressure but low temperature to prevent monosaccharide degradation. But in comparison to steam explosion and ammonia explosion processes, the sugar recovery yield from this process is lower. The carbon dioxide explosion pre-treatment of pure avicel (crystalline cellulose) was studied by Zheng *et al.*[48] This process was operated in a pressure range of 6.9–27.6 MPa at temperatures of 25, 35 and 80 °C. The pre-treatment enhanced the hydrolysis rate of crystalline cellulose. A similar result was also reported in supercritical carbon dioxide pre-treatment of corn stover.[49]

6.5.2.6   Organosolv Pre-treatment

The organosolv process is a process that uses an organic or aqueous organic solvent mixture (methanol, ethanol, acetone, ethylene glycol, tri-ethylene glycol and tetra-hydrofurfuryl alcohol) with an inorganic acid catalyst (HCl or H_2SO_4) or alkali catalyst to break down the lignin and hemicellulose structure. Temperatures used for the process can be as high as 200 °C, depending on the type of solvent and the structure of the biomass. Several investigations have been published for the pre-treatment of various biomass using organosolv process, including hardwood, grass and agricultural residues.[50–53]

6.5.3   Biological Pre-treatment

In a conventional bioethanol process, chemical and physicochemical methods are commonly used to pre-treat lignocellulosic raw materials for the subsequent enzymatic hydrolysis. Biological pre-treatment is however an alternative process of removing lignin and degrading hemicellulose and cellulose.[54–56] The biological pre-treatment process is becoming increasingly attractive owing to its mild conditions and is relatively safe and environmental benign in comparison to chemical and physicochemical pre-treatments. Moreover, biological pre-treatment does not generate toxic compounds, *e.g.* furfural and hydroxymethylfurfural (HMF), thus reducing the inhibitory effect of the lignocellulosic hydrolysate.

Many species of ligninolytic microorganisms have been investigated for the biological pre-treatment of lignocellulosic raw materials, including white-rot fungi, brown-rot, soft-rot fungi and cellulolytic bacteria.[57–60] Solid state fermentation is commonly used in conjunction with biological pre-treatment in a broad range of culture times ranged from 7 days to 150 days.[61,62]

Zeng *et al.* employed solid state fermentation of *Phanerochaete chrysosporium* on wheat straw which resulted in the degradation of 25% of the total lignin within 7 days, coupled with approximately 250% higher efficiency on the sugars released from wheat straw *via* enzymatic hydrolysis.[61] Salvachua *et al.* optimized the process parameters such as inoculum preparation, wheat straw particle size, moisture content, organic and inorganic supplementations, and mild alkali washing during solid-state fermentation, for the biological pre-treatment of wheat straw with the white-rot fungus *Irpex lacteus* and achieved glucose yields from wheat straw of 68% after 21 days as compared to yields of 62% and 33% for cultures grown without supplementation and on untreated raw material, respectively.[63]

In summary, currently there is no method that could meet all the requirements for the pre-treatment of lignocellulosic raw materials. Table 6.2 compares the advantages and disadvantages of various methods that are commonly used in the advanced generation bioethanol production process.

Table 6.2 Comparison of the advantages and disadvantages of various pretreatment methods.[33,41,64]

Pre-treatment process	Advantages	Disadvantages
Mechanical	Reduce cellulose crystallinity Increase surface area No inhibitor produced	High energy input
Dilute acid	Practical and simple technique Effectively degrade hemicelluloses Alter lignin structure	Inhibitors produced Corrosion of equipment
Concentrated acid	High glucose yield Completely degrade hemicelluloses	Inhibitors produced Expensive Corrosion of equipment Acid recycle required
Alkali	Remove hemicelluloses and lignin Increase surface area Low formation of inhibitors	Slow Complex inhibitors Less effective on softwoods
Steam explosion	Hemicellulose degradation and lignin transformation Fast Inexpensive	Inhibitors Acid catalysts required for biomass with high lignin content
Ammonia fibre explosion	Increases surface area Partial removal of hemicelluloses and lignin Low formation of inhibitors	Expensive Hazardous Not suitable for high lignin content Ammonia recycle required
Carbon dioxide explosion	Increases surface area Non hazard chemicals	Expensive High energy input
Organosolv	Hydrolyse lignin and hemicelluloses Possible to dissolve different biomass	Expensive Solvents need to be recovered and recycled
Biological pre-treatment	Degradation of hemicelluloses and lignin Low energy input Environmentally friendly No inhibitor generated	Slow Loss of cellulose

6.6   Detoxification

The detoxification step is only used when inhibitory compounds are formed, as in the case of acid pre-treatment process and alkali pre-treatment process. There are several kinds of methods that could be used to remove the inhibitory compounds in the hydrolysate, including neutralization, overliming with calcium hydroxide, activated charcoal, ion-exchange resins and enzymatic detoxification.[65] The selection of detoxification method depends on the property of the lignocellulose hydrolysate and in some cases a combination of two methods is required. For example, in the detoxification of a wheat straw hydrolysate from dilute acid pre-treatment, ion exchange-D 311 and overliming were used and removed 90.36% of the furfurals, 77.44% of the phenolic and 96.29% of the acetic acid produced.[66] In a lignocellulose to bioethanol project design published by the National Renewable Energy Laboratory (NREL), an ammonia detoxification method was proposed.[67] In this method, ammonia sulfate was added to neutralize the residue acid from the pre-treatment and also the inhibitors generated during the hydrolysis. The ammonia salts produced as the result of neutralization were then used as the nitrogen source for the growth of the microorganisms. Detoxification has been proven to improve the fermentability of various lignocellulose hydrolysates.

6.7   Hydrolysis

With many of the pre-treatment methods discussed above, only partial degradation of cellulose, hemicellulose and lignin is achieved. In order to fully utilize the sugars stored in the lignocellulosic raw materials, the hydrolysis step is indispensable. The hydrolysis process breaks down the hydrogen bonds in the cellulose and hemicellulose structures to liberate simple pentoses and hexoses. This process is commonly catalysed by various glucosly hydrolases, including mainly cellulase, hemicellulase and ligninase.

6.7.1   Glycosyl Hydrolases

6.7.1.1   Cellulase

Cellulase refers to a group of enzymes that hydrolyse cellulose fibres. Cellulolytic enzymes can be divided into the following three types:

1. Endoglucanases, also called endo-1,4-β-D-glucanase (EC 3.2.1.4), which hydrolyse internal β-1,4-D-glucosidic linkages randomly in the cellulose chain. Endoglucanases attack the low crystallinity regions of the cellulose creating free chain ends, which are commonly measured by detecting the reducing groups released from carboxymethylcellulose (CMC).
2. Exoglucanases (also known as cellobiohydrolases, CBH) (EC 3.2.1.91), which work on the crystalline cellulose by cutting cellobiose units from the free chain ends created by the endoglucanases. There are two main types of cellobiohydrolases. Cellobiohydrolase I (CBH I) functions on

the reducing end, and cellobiohydrolase II (CBH II) works on the non-reducing end of cellulose. Cellobiohydrolases are the only enzymes that efficiently degrade crystalline cellulose. However, cellobiohydrolases are inhibited by their hydrolysis product, cellobiose.
3. β-glucosidases (also known as cellobiase, EC 3.2.1.21), which releases D-glucose from cellobiose and soluble cellodextrins.

These three groups of enzymes work synergistically to degrade cellulose by creating new sites for each other and preventing product inhibition. The cellulase hydrolysis starts with endoglucanases which hydrolyse cellulose fibres into shorter chains of cellulose polymer, the action of exoglucanases reduces the degree of polymerization by cutting off a molecule of cellobiose from the cellulose chain, then β-glucosidases hydrolyse cellobiose into two molecules of glucose.

6.7.1.2   Hemicellulase

Hemicellulase is a group of enzymes that degrades hemicellulose into its corresponding monomer sugars. The main hemicellulase includes:

1. 1,4-β-D-xylan xylanohydrolase or endo-1,4-β-xylanase (EC.3.2.1.8), which breaks down xylosidic linkages in the xylan structure into β-D-xylopyranosyl oligomers and reduces the degree of polymerization.
2. 1,4-β-D-xylan xylohydrolase or 1,4-β-D-xylosidase (EC.3.2.1.37), which hydrolyses xylo-oligasaccharide and xylobiose to D-xylose residues.
3. Galactanase (EC 3.2.1.89), which hydrolyses 1,4-β-galactosidic linkages in type I arabinogalactans.
4. Arabinanase, which removes L-arabinose residues in the hemicellulose polymer. There are two types of arabinanases:
   a. exo α-L-arabinofuranosidase (EC 3.2.1.55)
   b. endo 1,5-L-arabinanase (EC 3.2.1.99)

6.7.1.3   Ligninase

Lignin degradation needs fungi and bacteria that produce ligninolytic enzymes or ligninase. Lignocellulolytic enzymes are oxidative enzymes which include:

1. laccases or para-diphenol oxygen oxidoreductase or phenol oxidase (EC 1.10.3.2). This is an enzyme that oxidizes aromatic amines and phenolic compounds
2. lignin peroxidase (LiP) (EC 1.11.1.14), another enzyme that oxidizes aromatic amines and phenolic compounds but only when veratryl alcohol is present
3. manganese peroxidase (MnP) (EC 1.11.1.13) that oxidizes aromatic amines and phenolic compounds which need manganese ion (Mn^{2+}) as the co-factor.

6.7.2 Enzyme Production

These cellulases and hemicellulases can be obtained from a wide variety of bacteria belonging to the family of *Clostridium, Cellulomonas, Bacillus, Thermomonospora, Ruminococcus, Bacteriodes, Erwinia, Acetovibrio, Microbispora* and *Streptomyces* as well as fungi belonging to the species of *Trichoderma, Aspergillus, Schizophyllum* and *Penicillium*.[68] Out of all of the fungi species capable of producing cellulases, *Trichoderma* has emerged as the most studied strain and was widely used for the production of cellulases in commercial scales. *T. reesei* releases a mixture of cellulases including five endo-glucanases and two exo-glucanases.[69] It however produces inefficient β-glucosidase, which is essential for the final stage of the hydrolysis process. Therefore, commercial cellulase products produced from *T. reesei* are often supplemented with extra β-glucosidases.[16] *Aspergillus niger*, as an alternative, is a very good producer of β-glucosidases and has been used for the production of cellulases as well.[16]

Trichoderma reesei also produces a range of hemicellulases. Being a diverse group of heterogeneous polymers with various side groups, the hemicellulosic system is more complex and involves several hemicellulases, for example xylanases, mannanases, arabinanases and various esterases. Depending on the substrate and pre-treatment, hemicellulases can be crucial for efficient hydrolysis of the cell wall.[69]

White rot fungi such as *Phanerochaete chrysosporium, Grammothele subargentea, Irpex lacteus, Trametes* versicolor, *Pleurotus ostreatus* present high lignocellulolytic activity. *Phanerochaete chrysosporium*, a model organism for lignin degradation study, produces high activity of manganese peroxidase (101 U/L) and lignin peroxidase (77 U/L).[70]

Cellulase can be produced by either solid-state fermentation (SSF) or submerged fermentation (SuF). Solid-state fermentation has some advantages of low energy consumption, low substrate cost, low operation cost and easy downstream processing. However, most of current commercial cellulase productions are carried out *via* submerged fermentation. This is due to the fact that it is easier to control the culture conditions in liquid fermentations. Table 6.3 gives a few examples of the glycosyl hydrolase production using agriculture waste materials.

6.8 Fermentation

6.8.1 Yeast Fermentation and Yeast Genetic Modification

After detoxification and enzymatic hydrolysis, the sugar solution is transferred into a fermenter for the bioethanol fermentation. Depending on the solid loading rate of the hydrolysis, the hydrolysate usually contains 60–100 g/L glucose. Certainly, a higher glucose concentration is desired as this would result in a higher bioethanol concentration in the fermentation beer. However, a solid loading rate in excess of 30% is usually extremely

Table 6.3 Cellulase, hemicellulase and ligninase production using agriculture waste materials in SSF and SuF.

Enzyme	Strain	Substrate	Fermentation	Activity	Ref.
Cellulase	T. harzianum T2008	Oil palm biomass	SSF	10.1 FPA U/g	71
Cellulase	T. harzianum T2008	Malt extract	SSF	8.2 FPA U/g	71
Cellulase	A. heteromorphus	Rice straw	SSF	14.1 FPA U/g	22
Cellulase	T. reesei RUT C-30	Cellulose –yeast extract	SuF	5.02 FPA U/ml	72
Cellulase	Acremonium cellulolyticus	Rice straw	SuF	11 FPA U/ml	73
Endoglucanase	A. heteromorphus	Rice straw	SSF	235.6 U/g	22
Endoglucanase	T. reesei	Soybean hulls and wheat bran	SSF	60.17 IU/g-ds	74
	A. oryzae			68.36 IU/g-ds	
	Co-culture of T. reesei and A. Oryzae			100.67 IU/g-ds	
β-glucosidase	T. reesei	Soybean hulls and wheat bran	SSF	6.30 IU/g-ds	74
	A. oryzae			9.45 IU/g-ds	
	Co-culture of T. reesei and A. Oryzae			10.71 IU/g-ds	
Xylanase	A. niger 3T5B8	Mango peel	SSF	20.33 U/ml	74
Xylanase	A. niger 3T5B8	Wheat bran	SSF	30.62 U/ml	74
Xylanase	Clostridium bifermentans LU-1	Oat spilt xylan	SuF	4.2 U/ml	75
Xylanase	T. reesei RUT C-30	Oat husk hydrolysate with lactose	SuF	1350 IU/ml	76
Polygalacturonase	A. niger 3T5B8	Wheat bran	SSF	30.75 U/ml	74
Laccase	Grammothele subargentea	Wood chips	SSF	0.292 U/ml	77

difficult to handle in terms of pumping and mixing and is therefore not favourable. Similar to the first generation bioethanol production process, simultaneous saccharification and fermentation attracts great interests in the advanced generation bioethanol production process. Instead of carrying out enzymatic hydrolysis in a separate vessel, glucosyl hydrolyases are directly added into the fermenter together with the fermentation microorganisms, *e.g. Saccharomyces cerevisiae* or *Zymomonas mobilis*. Simultaneous saccharification and fermentation has the advantages of reducing the inhibitory effects on both the enzymes and bioethanol producing strains, thereby increasing productivity and reducing operation cost.[78] However, the optimum temperatures for the hydrolysis and bioethanol fermentation are different, thus the hydrolysis efficiency is affected.

As cellulose only represents 30–40% of the lignocellulosic biomass, the utilization of only glucose in the fermentation would undeniably have an effect on the overall biomass to ethanol conversion yield. Therefore, it is necessary to consider the utilization of the hemicellulose hydrolysate in the development of the advanced generation bioethanol production process.[67,79]

One of the challenges of using hemicellulose hydrolysate is to identify a suitable microorganism that can ferment pentoses, the main monosaccharides of hemicellulose. The current industrial bioethanol-producing microorganisms, *e.g. S. cerevisiae*, convert glucose effectively and have high tolerance to ethanol. But they are unable to ferment pentoses. To overcome this setback, intensive research has been carried out to construct the xylose utilization metabolic pathways in *S. cerevisiae* strains. The genes *XYL1* (encoding xylose reductase, XR), *XYL2* (encoding xylitol dehydrogenase, XDH) and *XYL3* (encoding xylulokinase, XK) from *Pichia stipitis* and other natural xylose-fermenting yeasts have been expressed in *S. cerevisiae*.[80] The resulting recombinant yeasts were able to use xylose as the sole carbon source for the bioethanol production, though both the ethanol concentration and ethanol yield were low. It was reported that the redox imbalance was responsible for the low xylose to bioethanol efficiency, as the XR utilized NADPH as the coenzyme while the XDH was strictly NAD dependent.[81] Several attempts have been made to express an NADH-dependent XR in the *S. cerevisiae*, leading to an increased ethanol yield of 0.46 g/g in laboratory fermentations at best conditions.[80,82] The research in this area has been recently reviewed by Kim *et al.*[83] Besides xylose, recombinant yeasts that can ferment arabinose have also been constructed, though the ethanol yield obtained was only 0.39 g ethanol per g consumed arabinose with a very low production rate of 0.036 g ethanol per g cell dry weight.[84]

6.8.2 Yeast and Tolerance to Inhibitors

Another challenge in the development of the advanced generation bioethanol production is the presence of inhibitors in the hydrolysate. Using dilute acid pre-treatment as an example, the common inhibitors detected in the

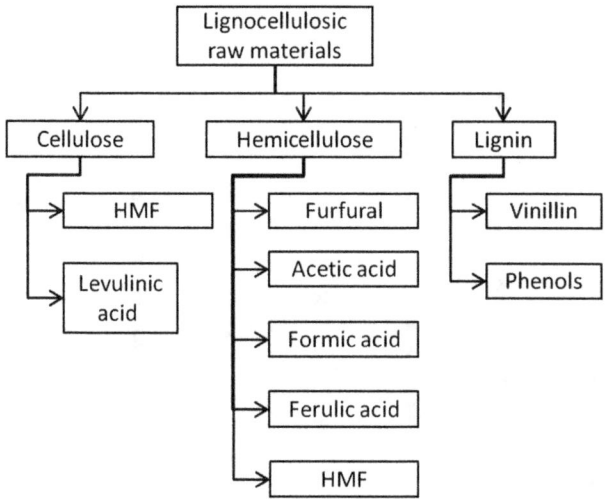

Figure 6.3 Inhibitors commonly detected in the hydrolysate of lignocellulosic raw materials.

hydrolysate include acetic acid, formic acid, furfural, hydroxymethylfurfural (HMF), vanillin and phenols. These inhibitors extend the lag phase, reduce the ethanol yield and decrease (and even completely suppress) the cell growth. Figure 6.3 shows the corresponding components that these inhibitors are derived from.

The composition and concentration of the inhibitors are dependent on the biomass substrate, pre-treatment method and pre-treatment condition. Tomas-Pejo *et al.* reported that in a steam explosion pre-treated wheat straw hydrolysate, furfural, HMF, acetic acid, formic acid and ferulic acid were detected with concentrations of 1.4, 0.1, 5.1, 1.3 and <0.1 g L^{-1}, respectively.[85] Figure 6.4 summarizes the common inhibitor concentrations based on the data reported in two recent review papers.[66,86] The furfural concentration was found to be in the range of 0.15 to 2.2 g L^{-1}, while the HMF was generally lower than 1 g L^{-1} except for spruce-derived hydrolysates. Acetic acid and formic acid concentrations were around 1.6 to 5.5 g L^{-1} and 0.7 to 3.1 g L^{-1}, respectively. Phenolics had a wide spread from 0.1 to 4.5 g L^{-1}, depending on the pre-treatment and detection methods. Only few publications reported levulinic acid inhibitor, which is mainly from woody biomass.

The critical concentrations of the inhibitor were dependent on the microorganisms used in the ethanol fermentation. Yeast strains have been proven to be more tolerant than bacterial bioethanol producers (*e.g. Zymomnas mobilis* and recombinant *E. coli*).[86] Martin and Jonsson reported that 7.5 g L^{-1} HMF or 3.0 g L^{-1} furfural or 1.2 g L^{-1} HMF plus 1.0 g L^{-1} furfural extended the lag phase of *S. cerevisiae* ATCC 211239 over 24 hours.[87] When the HMF or furfural concentration was higher than 15.0 and

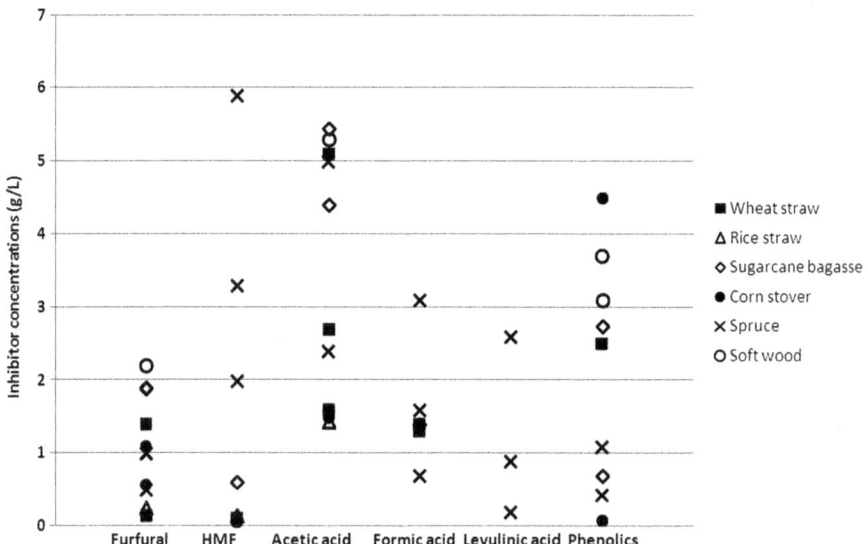

Figure 6.4   Inhibitor concentrations detected in the hydrolysates of various bio-mass, data obtained from Almeida *et al.*, Chandel *et al.* and Tomas-Pejo *et al.*[66,85,86]

6.0 g L^{-1} respectively, the *S. cerevisiae* growth was completely suppressed. Several mechanisms have been proposed for the inhibitory effect caused by furans. Direct inhibition of alcohol dehydrogenase (ADH), pyruvate dehydrogenase (PDH) and aldehyde dehydrogenase (ALDH) were observed by Modig *et al.*[88] The reduction of furans consumes NAD(P)H, leading to an imbalance of the intracellular redox potential. Furfural also affects the glycolysis pathway and TCA cycle; such alters the ATP formation.[89]

The weak organic acids such as acetic acid and formic acid both have positive and negative effects on the bioethanol production process. In fermentations using *S. cerevisiae* NCYC 2592, an addition of acetic acid in a concentration of 20 mM increased the ethanol productivity (unpublished data). The low acetic acid concentrations (lower than 20 mM) did not have an impact on the yeast viability. At fermentations with higher acid concentration, the intracellular pH decreases, requiring plasma membrane ATPase to pump protons out of the cell.[86] The depletion of ATP affected the biomass formation. In comparison with acetic acid, formic acid has a more severe inhibitory effect, which has also been observed in other biosynthesis processes, *e.g.* succinic acid formation.[90]

6.8.2.1   High Tolerance Strains

Besides detoxification, another promising approach to overcome the inhibitory effect is to develop high tolerance strains. A robust industrial strain should not only tolerate high levels of inhibitors present in the hydrolysate, but also high ethanol concentration, high osmosis and, ideally, high temperature. Strain adaptation is a common and effective method for the screening of high inhibitor tolerance strains. Several publications have reported improved lignocellulosic to bioethanol fermentation using adapted yeast strains.[91–93] In a repetitive batch culture of 428 generations of *S. cerevisiae* in a medium containing a cocktail of 12 inhibitors, the yeast strain improved its maximum specific growth rate from 0.18 h^{-1} to 0.33 h^{-1} and the lag phase decreased from 48 h to 24 h. In a chemostat adaptation after 97 generations, the isolated yeast strains showed a 4-fold higher specific furfural conversion rate and up to 50% higher specific ethanol productivity.[93] Strain hybridization has also been investigated to generate high tolerance yeast strains. Benjaphokee *et al.* hybridized a high temperature tolerant *S. cerevisiae* with a high ethanol productivity strain.[94] The resultant hybrid *S. cerevisiae* TJ14 exhibited high ethanol producing capacity at high temperature (41 °C), together with high acid tolerance.

Genetic modification of the yeast strain has also been investigated to improve bioethanol production using the medium containing fermentation inhibitors.[95–97] Over-expression of the genes encoding an NADPH-dependent alcohol dehydrogenase led to a yeast strain with over 4-fold increase in HMF consumption.[98]

6.8.3 Co-fermentation using Yeast Strains

Another strategy of utilizing both hexoses and pentoses is to culture two yeast strains simultaneously in which one strain (*e.g. S. cerevisiae*) is capable of fermenting glucose while the other (*e.g. P. stipitis*) is capable of fermenting xylose. Table 6.4 summarizes briefly the advantages and disadvantages of fermentations using *S. cerevisiae* or *P. stipitis* alone. The main disadvantage of *S. cerevisiae* is that it cannot ferment C5 sugars. While co-culturing of a *P. stipitis* strain with *S. cerevisiae* has the potential to overcome this shortcoming, Wan *et al.* co-cultured *S. cerevisiae* Y5 and *P. stipitis* CBS6054 on non-detoxified dilute acid hydrolysate.[99] The results showed that both xylose and glucose were effectively converted to ethanol with an overall ethanol yield of 85% of the theoretical yield. Moreover, both these strains exhibited good tolerance to the inhibitors furfural and HMF, by converting them to less toxic compounds. Similarly, co-fermentation of *S. cerevisiae* with *Spathaspora arborariae* was carried out using rice hull hydrolysate.[100] Around 14.5 g L^{-1} ethanol and 3 g L^{-1} xylitol were obtained from the glucose and xylose presented in the hydrolysate. Chandel *et al.* compared monocultures and co-culture of a *P. stipitis* strain and a thermotolerant *S. cerevisiae* strain on an acidic hydrolysate of sugarcane bagasse.[101] The co-culture resulted in an ethanol concentration of 15.0 g L^{-1}, which was higher than both the mono culture of the *P. stipitis* (1.40 g L^{-1}) and the mono culture of the *S. cerevisiae* (12.1 g L^{-1}).

As the two microorganisms may have different optimum pH, temperature, oxygen demand and nutrient, the optimization of the fermentation involving

Table 6.4   Advantages and Disadvantages of Fermentations Using *S. Cerevisiae* and *P. Stipitis.*[8]

Strains	Advantages	Disadvantages
S. cerevisiae	• Naturally adapted to ethanol fermentation • High ethanol yield • High tolerance to ethanol and inhibitors • Amenability to genetic modifications	• Unable to ferment xylose and arabinose
P. stipitis	• Good xylose fermentation • Good ethanol yield • Could ferment most of cellulosic-material sugars including glucose, galactose and cellobiose	• Low ethanol tolerance • Sensitive to inhibitors • Requires micro-aerobic conditions to reach peak performance

Table 6.5   Bioethanol production from fermentations using immobilized *S. cerevisiae* in various entrapment matrix.

Strains	Immobilization matrix	Ethanol $(g\ L^{-1})$	Yield (%)	Productivity $(g\ L^{-1}\ h^{-1})$	Operation time (days)	Ref.
S. cerevisiae	Carrageen and silica	42	–	–	105	103
S. cerevisiae	Calcium alginate	57	97	–	11	105
S. cerevisiae	Calcium alginate	85	95	25	>90	104
S. cerevisiae	Sponge	24	–	7.1	–	106

two strains is challenging. One possible solution is to immobilize one or two strains within a porous matrix. Although some studies indicated that the immobilization did not improve the ethanol production, the immobilized cells had an increased tolerance to higher substrate and product concentrations compared to that of the free cells.[102] This is of particular interest to fermentations using lignocellulosic hydrolysate.

Many studies have been carried out using entrapment as the immobilization technique and Table 6.5 summarizes some early studies using immobilized *S. cerevisiae* for bioethanol production. A steady high-level bioreactor performance was achieved in studies reported by Karkare *et al.* using carrageen and silica as the immobilizer and by Nagashima *et al.* using calcium alginate as the immobilizer.[103,104] The latter achieved the highest ethanol concentration and a 95% yield along with a relatively high productivity of 25 g L^{-1} h^{-1}. McGhee *et al.* also reported a high bioethanol yield (97%) in a fermentation using calcium alginate immobilized yeast.[105] Del Borghi *et al.* showed that *S. cerevisiae* entrapped in a sponge matrix has a much lower productivity than that obtained using the calcium alginate,[106] and the ethanol concentration achieved in this study was 24 g L^{-1}, suggesting that sponge was not as viable a matrix as that of calcium alginate.

Isabella *et al.* co-cultured *S. cerevisiae* with *Scheffersomyces stipitis* strain NRRLY-11544 in liquid fermentations containing both glucose and xylose.[107] The results showed that the co-cultures resulted in faster processes than single cultures of *S. Stipitis*. However, the high ethanol production by *S. cerevisiae* inhibited *S. stipitis*. In another approach, Silva *et al.* immobilized glucose isomerase (GI) in chitosan, then the support containing the immobilized GI was co-immobilized with *S. cerevisiae* in alginate gel.[108] In this system, GI converted xylose to xylulose and then the *S. cerevisiae* fermented xylulose to bioethanol and other products. In a fermentation using a medium containing 50 g L^{-1} initial xylose, 12 g L^{-1} ethanol, 9.5 g L^{-1} xylitol, 2.5 g L^{-1} glycerol and 1.9 g L^{-1} acetate were produced in 48 h.

6.8.4 Thermophilic Bioethanol Production

Besides yeast strains, many type of other microorganisms could also be used for bioethanol production, such as *Clostridium* sp. and some thermophilic strains.[109] Conventional bioethanol fermentations by yeast strains are carried out at mesophilic temperatures, *e.g.* 30 °C. There are, however, some bacteria that can grow and synthesize ethanol at relatively higher temperatures, *e.g.* 70 °C. These thermophilic microorganisms include *Geobacillus stearothermophillus*, *Geobacillus thermoglucosidasius*, *Thermoanerobacterium saccharolyticum*, *Thermoanerobacter mathranii*, *Clostridium thermocellum*, *Thermoanerobacterium thermosaccharolyticum* and *Thermoanaerobacter thermohydrosulfuricus*.

Natural *S. cerevisiae* strains can only use C6 sugars for the bioethanol fermentation, but cannot utilize C5 sugars. There are, however, some thermopholic bioethanol producing strains, *e.g. Geobacillus stearothermophillus*, *Geobacillus thermoglucosidasius*, which can utilize either glucose or xylose/ arabinose as the substrate. This allows for a wider choice of 'fermentable carbohydrates' that could not be accommodated in the first generation bioethanol production processes. At high temperatures, the saturated ethanol concentration in the fermentation broth is lower than that at low temperatures. The high temperature causes the ethanol to evaporate thereby reducing the ethanol inhibition to microorganisms. Moreover, the high temperature prevents contamination from microorganisms that cannot tolerate this temperature. But contamination can still happen due to the existence of un-expected thermopholic strains.

Geobacillus stearothermophillus, for example is one of the most commonly used thermophilic bioethanol producing strains. It is a rod-shaped, Grampositive bacterium with a living temperature of 30–75 °C. It was first isolated from food waste materials. Wild *G. stearothermophillus* strain produces lactic acid as the main product with trace amounts of ethanol, acetic acid and formic acid as by-products. The genetic modification on *G. stearothermophillus* has been investigated to block the pathway from pyruvate to

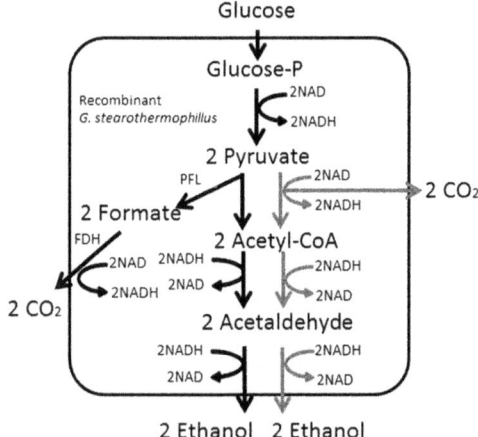

Figure 6.5 An example of the metabolic pathway of a recombinant *G. stearother-mophillus*. The pathway from pyruvate to lactic acid was deleted and the gene encoded pyruvate formate lyase (PFL) was over-expressed. FDH: formate dehydrogenase.[110]

lactic acid, leading to the accumulation of ethanol as the main product. Figure 6.5 shows a typical recombinant strain of *G. stearothermophillus*, in which the lactate dehydrogenase gene was deleted while a NAD-linked formate dehydrogenase gene was introduced. The lactic acid producing pathway has been replaced by PFL (pyruvate formate lyase) pathway with the formate being converted to NADH and CO_2 by the enhanced formate dehydrogenase activity. The theoretical glucose to ethanol yield for the recombinant strain is 2.

6.9 Conclusion

In the past decade, bioethanol production expanded dramatically world-wide. Petroleum with added bioethanol has been widely used in the transportation sector in US, Brazil and many European countries. Currently, bioethanol is predominately produced *via* the first generation production process, using food-derived materials as the substrate. With the increasing concern about global food security, the conversion of food-based materials to bioethanol will have to face escalating challenges. Therefore, most of the investigations in bioethanol synthesis have turned their focus on bioethanol production from non-food materials. Various lignocellulosic raw materials have been tested and different pre-treatment methods have been proposed, aiming to reduce the cost of generating a fermentable sugar solution. However, there is still no clear conclusion about which substrate, which pre-treatment method or which combination of substrate and pre-treatment method is the best. Similarly, a wide range of strains, including *S. cerevisiae*, *P. stipitis*, *Clostridium* sp. and thermophiles have been intensively studied for fermenting the lignocellulosic hydrolysate into ethanol. Although further research is required to overcome the hindrances, the latest progress in this field shows promising improvements toward the development of an economic feasible process for the advanced generation of bioethanol production.

References

1. BP, *BP Energy Outlook 2030*, 60 years BP Statistical Review, 2011, pp. 4–80.
2. International Energy Agency, *Technology Roadmap: Biofuels for Transport*, France, 2011, pp. 1–51.
3. N. Pensupa, M. Jin, M. Kokolski, D. B. Archer and C. Du, *Bioresour. Technol.*, 2013, **149**, 261.
4. M. O. S. Dias, A. V. Ensinas, S. A. Nebra, R. Maciel Filho, C. E. V. Rossell and M. R. W. Maciel, *Chem. Eng. Res. Design*, 2009, **87**, 1206.
5. A. Koutinas, C. Du, C. S. K. Lin and C. Webb, *Advances in Biorefineries: Biomass and Waste Supply Chain Exploitation*, ed. K. W. Waldron, Woodhead Publishing, UK, 2014.
6. S. Patumsawad, *J. Sustain. Energy Environ. Special Issue*, 2011, **2011**, 47.
7. R. Ibbett, S. Gaddipati, S. Davies, S. Hill and G. Tucker, *Bioresour. Technol.*, 2011, **102**, 9272.
8. A. Limayem and S. C. Ricke, *Prog. Energ. Combust. Sci.*, 2012, **38**, 449.
9. A. J. A. van Maris, D. A. Abbott, E. Bellissimi, J. van den Brink, M. Kuyper, M. A. H. Luttik, H. W. Wisselink, W. A. Scheffers, J. P. van Dijken and J. T. Pronk, *Antoine van Leeuwenhoek*, 2006, **90**, 391.
10. D. J. Cosgrove, *Nature Rev. Mol. Cell Biol.*, 2005, **6**, 850.
11. M. T. García-cubero, M. Coca, S. Bolado and G. Gonzalez-Benito, *Chem. Eng. Trans.*, 2010, **21**, 1273.
12. S. I. Mussatto and J. A. Teixeira, *Curr. Res., Technol. Educ. Top. Appl. Microbiol. Microb. Biotechnol.*, 2010, 897.
13. R. R. Singhania, B. Parameswaran and A. Pandey, *Handbook of Plant-Based Biofuels*, ed. A. Pandey, CRC Press, 2008.
14. P. Sannigrahi, A. J. Ragauskas and G. A. Tuskan, *Biofuels, Bioproducts Biorefining*, 2010, **4**, 209.
15. P. F. H. Harmsen, W. J. J. Huijgen, L. M. Bermudez Lopez and R. R. C. Bakker, *Literature Review of Physical and Chemical Pre-treatment Processes for Lignocellulosic Biomass*, Energy Research Centre of the Netherlands, *Netherlands*, 2010, pp. 1–33.
16. M. J. Taherzadeh and K. Karimi, *Bioresour. Technol.*, 2008, **2**, 707.
17. A. Sant'Ana da Silva, H. Inoue, T. Endo, S. Yano and E. P. Bon, *Bioresour. Technol.*, 2010, **101**, 7402.
18. A. M. Elshafei, *Bioresour. Technol.*, 1991, **35**, 73.
19. D. B. Rivers and G. H. Emert, *Biol. Wastes*, 1988, **26**, 85.
20. K. L. Chang, J. Thitikorn-amorn, J. F. Hsieh, B. M. Ou, S. H. Chen, K. Ratanakhanokchai, P. J. Huang and S. T. Chen, *Biomass Bioenergy*, 2011, **35**, 90.

21. C. Yanga, Z. Shena, G. Yub and J. Wang, *Bioresour. Technol.*, 2008, **99**, 6240.
22. A. Singh, S. Tuteja, N. Singh and N. R. Bishnoi, *Bioresour. Technol.*, 2011, **102**, 1773.
23. L. Z. Xin and M. Kumakura, *Bioresour. Technol.*, 1993, **43**, 13.
24. D. Christiane, A. Daro and D. Monnoye, *Eur. Polym. J.*, 1980, **16**, 1159.
25. C. Cara, E. Ruiz, J. Olival, F. Sáez and E. Castro, *Bioresour. Technol.*, 2008, **99**, 1869.
26. S. Kim, S. Lee, E. Jang, S. Han, C. Park and S. Kim, *J. Ind. Eng. Chem.*, 2011, **18**, 183.
27. J. R. Jensen, J. E. Morinelly, K. R. Gossen, M. J. Brodeur-Campbell and D. R. Shonnard, *Bioresour. Technol.*, 2010, **101**, 2317.
28. G. L. Guo, W. H. Chen, W. H. Chen, L. C. Men and W. S. Hwang, *Bioresour. Technol.*, 2008, **99**, 6046.
29. T. C. Hsu, G. L. Guo, W. H. Chen and W. S. Hwang, *Bioresour. Technol.*, 2010, **101**, 4907.
30. B. Y. Cai, J. P. Ge, H. Z. Ling, K. K. Cheng and W. X. Ping, *Biomass Bioenergy*, 2012, **36**, 250.
31. J. Zhang, B. Zhang, J. Zhang, L. Lin, S. Liu and P. Ouyang, *Biotechnol. Adv.*, 2010, **28**, 613.
32. P. Kumar, D. M. Barrett, M. J. Delwiche and P. Stroeve, *Ind. Eng. Chem. Res.*, 2009, **48**, 3713.
33. G. Brodeur, E. Yau, K. Badal, J. Collier, K. B. Ramachandran and S. Ramakrishnan, *Enzyme Res.*, 2011, **2011**, 1.
34. Y. H. Jung, I. J. Kim, J. I. Han, I. G. Choi and K. H. Kim, *Bioresour. Technol.*, 2011, **102**, 9806.
35. T. H. Kim, F. Taylor and K. B. Hicks, *Bioresour. Technol.*, 2008, **99**, 5694.
36. S. McIntosh and T. Vancov, *Bioresour. Technol.*, 2010, **101**, 6718.
37. B. C. Saha and M. A. Cotta, *Enzyme Microb. Technol.*, 2007, **41**, 528.
38. T. A. D. Nguyen, K. R. Kim, S. J. Han, H. Y. Cho, J. W. Kim, S. M. Park, J. C. Park and S. J. Sim, *Bioresour. Technol.*, 2010, **101**, 7432.
39. G. Chen and H. Chen, *Food Chem.*, 2011, **126**, 1934.
40. Y. Zheng, Z. Pan and R. Zhang, *Int. J. Agric. Biol. Eng.*, 2009, **2**, 51.
41. F. Carvalheiro, L. C. Duarte and F. M. Gírio, *J. Sci. Ind. Res.*, 2008, **67**, 849.
42. J. M. Lee, H. Jameel and R. A. Venditti, *Bioresour. Technol.*, 2010, **101**, 5449.
43. F. Teymouri, L. Laureano-Perez, H. Alizadeh and B. E. Dale, *Bioresour. Technol.*, 2005, **96**, 2014.
44. J. D. Blasig, M. T. Holtzapple, B. E. Dale, C. R. Engler and F. M. Byers, *Resour., Conservation Recycling*, 1992, 7, 95.
45. B. Z. Li, V. Balan, Y. J. Yuan and B. E. Dale, *Bioresour. Technol.*, 2010, **101**, 1285.
46. V. Balan, L. Sousa, S. P. Chundawat, R. Vismeh, D. A. Jones and B. E. Dale, *J. Ind. Microb. Biotechnol.*, 2008, **35**, 293–301.

47. H. Alizadeh, F. Teymouri, T. I. Gilbert and B. E. Dale, *Appl. Biochem. Biotechnol.*, 2005, **124**, 1133.
48. Y. Zheng, H. M. Lin and G. T. Tsao, *Biotechnol. Progr.*, 1998, **14**, 890.
49. N. Narayanaswamy, A. Faik, D. J. Goetz and T. Gu, *Bioresour. Technol.*, 2011, **102**, 6995.
50. N. Park, H. Y. Kim, B. W. Koo, H. Yeo and I. G. Choi, *Bioresour. Technol.*, 2010, **101**, 7046.
51. P. Sannigrahi, S. J. Miller and A. J. Ragauskas, *Carbohydr. Res.*, 2010, **345**, 965.
52. P. Obama, G. Ricochon, L. Muniglia and N. Brosse, *Bioresour. Technol.*, 2012, **112**, 156.
53. D. Gong, K. M. Holtman, D. Franqui-Espiet, W. J. Orts and R. Zhao, *Biomass Bioenergy*, 2011, **35**, 4435.
54. J. Shi, M. S. Chinn and R. R. Sharma-Shivappa, *Bioresour. Technol.*, 2008, **99**, 6556.
55. F. H. Sun, J. Li, Y. X. Yuan, Z. Y. Yan and X. F. Liu, *Int. Biodeterioration Biodegradation*, 2011, **65**, 931.
56. M. Taniguchi, H. Suzuki, D. Watanabe, K. Sakai, K. Hoshino and T. Tanaka, *J. Biosci. Bioeng.*, 2005, **100**, 637.
57. X. Yang, F. Ma, H. Yu, X. Zhang and S. Chen, *Bioresour. Technol.*, 2011, **102**, 3498.
58. M. J. Ray, D. J. Leak, P. D. Spanu and R. J. Murphy, *Biomass Bioenergy*, 2010, **34**, 1257.
59. K. Gopalakrishnan, J. v. Leeuwen and R. C. Brown, *Sustainable Bioenergy and Bioproducts*, Springer, London, 2012.
60. M. Dashtban, H. Schraft and W. Qin, *Int. J. Biol. Sci.*, 2009, **5**, 578.
61. J. Zeng, D. Singh and S. Chen, *Bioresour. Technol.*, 2011, **102**, 3206.
62. C. Xu, F. Ma and X. Zhang, *J. Agric. Food Chem.*, 2010, **58**, 10893.
63. D. Salvachúa, A. Prieto, M. E. Vaquero, Á. T. Martínez and M. J. Martínez, *Bioresour. Technol.*, 2013, **131**, 218.
64. D. Salvachúa, A. Prieto, M. López-Abelairas, T. Lu-Chau, Á. T. Martínez and M. J. Martínez, *Bioresour. Technol.*, 2011, **102**, 7500.
65. A. K. Chandel, R. K. Kapoor, A. Singh and R. C. Kuhad, *Bioresour. Technol.*, 2007, **98**, 1947.
66. A. K. Chandel, S. da Silva and O. Singh, *Biofuel Production – Recent Developments and Prospects*, ed. M. A. Dos Santos Bernardes, InTech, 2011, Chapter 10, pp. 1–23.
67. D. Humbird, R. Davis, L. Tao, C. Kinchin, D. Hsu, A. Aden, P. Schoen, J. Lukas, B. Olthof, M. Worley, D. Sexton and D. Dudgeon, *Process Design and Economics for Biochemical Conversion of Lignocellulosic Biomass to Ethanol: Dilute Acid Pretreatment and Enzymatic Hydrolysis of Corn Stover*, 2011, Report No. TP-5100-47764, pp. 1–114.
68. Y. Sun and J. Y. Cheng, *Bioresour. Technol.*, 2002, **83**, 1.
69. J. B. Kristensen, *Enzymatic Hydrolysis of Lignocellulose. Substrate Interactions and High Solids Loadings*, Prinfo Aalborg, Denmark, 2009, pp. 1–111.

70. C. Novotný, K. Svobodová, P. Erbanová, T. Cajthaml, A. Kasinath, E. Lang and V. Šašek, *Soil Biol. Biochem.*, 2004, **36**, 1545.
71. M. Z. Alam, A. A. Mamun, I. Y. Qudsieh, S. A. Muyibi, H. M. Salleh and N. M. Omar, *Biochem. Eng. J.*, 2009, **46**, 61.
72. A. Ahamed and P. Vermette, *Biochem. Eng. J.*, 2008, **40**, 399.
73. A. Hideno, H. Inoue, K. Tsukahara, S. Yano, X. Fang, T. Endo and S. Sawayama, *Enzyme Microb. Technol.*, 2011, **48**, 162.
74. S. Couri, S. da Costa Terzi, G. A. Saavedra Pinto, S. Pereira Freitas and A. C. Augusto da Costa, *Process Biochem.*, 2000, **36**, 255.
75. S. Marichamy and B. Mattiasson, *Enzyme Microb. Technol.*, 2005, **37**, 497.
76. H. Xiong, *Production and Characterization of Trichoderma reesei and Thermomyces lanuginosus Xylanase*, Helsinki University of Technology, Espoo, Finland, 2004, pp. 1–40.
77. M. C. N. Saparrat, P. Mocchiutti, C. S. Liggieri, M. B. Aulicino, N. O. Caffini, P. A. Balatti and M. J. Martínez, *Process Biochem.*, 2008, **43**, 368.
78. I. N. Ahmed, P. L. T. Nguyen, L. H. Huynh, S. Ismadji and Y. H. Ju, *Bioresour. Technol.*, 2013, **136**, 213.
79. A. Aden, M. Ruth, K. Ibsen, J. Jechura, K. Neeves, J. Sheehan, B. Wallace, L. Montague, A. Slayton and J. Lukas, *Lignocellulosic Biomass to Ethanol Process Design and Economics Utilizing Co-Current Dilute Acid Prehydrolysis and Enzymatic Hydrolysis for Corn Stover*, 2002, Report No. NREL/TP-510-32438, pp. 1–83.
80. A. Matsushika, H. Inoue, T. Kodaki and S. Sawayama, *Appl. Microbiol. Biotechnol.*, 2009, **84**, 37.
81. P. Kötter and M. Ciriacy, *Appl. Microbiol. Biotechnol.*, 1993, **38**, 776.
82. S. Watanabe, A. Abu Saleh, S. P. Pack, N. Annaluru, T. Kodaki and K. Makino, *Microbiology*, 2007, **153**, 3044.
83. S. R. Kim, Y. C. Park, Y. S. Jin and J. H. Seo, *Biotechnol. Adv.*, 2013, **31**, 851.
84. J. Becker and E. Boles, *Appl. Environ. Microbiol.*, 2003, **69**, 4144.
85. E. Tomas-Pejo, J. M. Oliva, M. Ballesteros and L. Olsson, *Biotechnol. Bioeng.*, 2008, **100**, 1123.
86. J. R. M. Almeida, T. Modig, A. Petersson, B. Hahn-Hagerdal, G. Lidén and M. F. Gorwa-Grauslund, *J. Chem. Technol. Biotechnol.*, 2007, **82**, 340.
87. C. Martin and L. J. Jonsson, *Enzyme Microb. Technol.*, 2003, **32**, 386.
88. T. Modig, G. Liden and M. J. Taherzadeh, *Biochem. J.*, 2002, **363**, 769.
89. I. Sarvari Horvath, C. J. Franzen, M. J. Taherzadeh, C. Niklasson and G. Liden, *Appl. Environ. Microbiol.*, 2003, **69**, 4076.
90. S. K. C Lin, C. Du, A. Koutinas, R. Wang and C. Webb, *Biochem. Eng. J.*, 2008, **41**, 128.
91. C. Fan, K. Qi, X. X. Xia and J. J. Zhong, *Bioresour. Technol.*, 2013, **136**, 309.
92. R. Landaeta, G. Aroca, F. Acevedo, J. A. Teixeira and S. I. Mussatto, *Appl. Energy*, 2013, **102**, 124.

93. R. Koppram, E. Albers and L. Olsson, *Biotechnol. Biofuels*, 2012, **5**, 1.
94. S. Benjaphokee, D. Hasegawa, D. Yokota, T. Asvarak, C. Auesukaree, M. Sugiyama, Y. Kaneko, C. Boonchird and S. Harashima, *New Biotechnol.*, 2012, **29**, 379.
95. E. Nevoigt, *Microbiol. Mol. Biol. Rev.*, 2008, **72**, 379.
96. D. Q. Zheng, X. C. Wu, X. L. Tao, P. M. Wang, P. Li, X. Q. Chi, Y. D. Li, Q. F. Yan and Y. H. Zhao, *Bioresour. Technol.*, 2011, **102**, 3020.
97. L. N. Jayakody, K. Horie, N. Hayashi and H. Kitagaki, *Appl. Microbiol. Biotechnol.*, 2013, **97**, 6589.
98. A. Petersson, J. R. Almeida, T. Modig, K. Karhumaa, B. Hahn-Hägerdal, M. F. Gorwa-Grauslund and G. Lidén, *Yeast*, 2006, **23**, 455.
99. P. Wan, D. Zhai, Z. Wang, X. Yang and S. Tian, *Biotechnol. Res. Int.*, 2012, **2012**, 1.
100. L. R. Hickert, F. da Cunha-Pereira, P. B. de Souza-Cruz, C. A. Rosa and M. A. Ayub, *Bioresour. Technol.*, 2013, **131**, 508.
101. A. K. Chandel, O. V. Singh, M. L. Narasu and L. V. Rao, *Biotechnology*, 2011, **28**, 593.
102. S. Nikolic, L. Mojovic, D. Pejin, M. Rakin and M. Vukasinovic, *Biotechnol. Bioeng.*, 2010, **34**, 1449.
103. S. B. Karkare, R. C. Dean Jr and K. Venkatasubramanian, *Nature Biotechnol.*, 1985, **3**, 247.
104. M. Nagashima, M. Azuma, S. Noguchi and K. Inzuka, *Arabian J. Sci. Eng.*, 1984, **2**, 299.
105. J. E. McGee, M. E. Carr and G. St.Julien, *Cereal Chem.*, 1984, **61**, 446.
106. M. Del Borghi, A. Converti, F. Parisi and G. Ferraiolo, *Biotechnol. Bioeng.*, 1985, **27**, 761.
107. D. B. Isabella, D. C. Paola, C. Daniela, L. Federico, C. Angela and R. Patrizia, *New Biotechnol.*, 2013, **30**, 591.
108. C. R. Silva, T. C. Zangirolami, J. P. Rodrigues, K. Matugi, R. C. Giordano and R. L. C. Giordano, *Enzyme Microb. Technol.*, 2012, **50**, 35.
109. A. L. Demain, M. Newcomb and J. H. D. Wu, *Microbiol. Mol. Biol. Rev.*, 2005, **69**, 124.
110. M. Javed and N. Baghaei-Yazdi, *Enhancement of Microbial Ethanol Production*, Bioconversion Technologies Limited, 2007, Patent, WO 2007-110606.

CHAPTER 7

High Value Chemicals and Materials Production Based on Biomass Components Separation

JIE CHANG

School of Chemistry and Chemical Engineering, South China University of Technology, No.381 Wushan Rd., Guangzhou, China, 510641
Email: changjie@scut.edu.cn

7.1 Introduction

With the development of global industrialization, the energy crisis and the environmental problems resulting from fossil resource use are becoming increasingly acute. Sustainable and environmentally friendly development has become a key issue in this century. As an alternative source of chemicals, materials, and energy, biomass has been becoming one of the most important sustainable sources due to its abundance and renewability.

Lignocellulosic biomass as the main bio-source consists of three main constituents in varying ratios: cellulose, lignin, and hemicellulose. Cellulose is the major structural polymer of a plant cell wall and usually exists as long thread-like fibers. It is a linear polysaccharide consisting of monomeric units of anhydro-D-glucose units with a β-1,4-glycosidic linkage.[1,2] Hemicellulose is a branched polysaccharide comprised of various sugar monomers including glucose, xylose, mannose, galactose, arabinose and uronic acids, and these

RSC Green Chemistry No. 27
Renewable Resources for Biorefineries
Edited by Carol Sze Ki Lin and Rafael Luque
© The Royal Society of Chemistry 2014
Published by the Royal Society of Chemistry, www.rsc.org

monomers can form hydrogen bonds with the cellulose and lignin.[2,3] Lignin is a phenolic high molecular mass biopolymer (600–15000 kDa), composed of a highly branched phenylpropanoid framework.[4–6] Crude lignin can be obtained in large quantities in the pulp and paper industry, mainly as kraft lignin and lignosulfonate. Moreover, lignin is a low value compound and has so far mainly been used as an energy source in combustion applications, with less than 5% being processed for other purposes at 2008.[7] Unlike lignin and hemicellulose, cellulose is the only biomass constituent used in a large-scale industrialized process, mainly producing paperboard and paper. However, according to a large number of form researches, all of these three constituents seem to have great potential for high-value industrial application due to their special structure. For example, lignin would be a sustainable source of phenolic compounds,[7,8] hemicelluloses can be used to synthesize sugars by hydrolysis,[9] while cellulosic ethanol is under investigation as an alternative fuel source.[10] Besides, all of the three main kinds of biomass components can be converted to bio-oils; however, the reaction pathways and conditions, quality and recovery yield of bio-oil would be different. Separation of the three main biomass components from each other for special study and application is important. In this chapter, high value chemicals and materials production based on separated biomass components is introduced and discussed.

7.2 Separating the Components of Biomass

A vast carbon source for bio-based products is locked up in plant matter, the most abundant source of biomass on earth. The principal components of biomass are cellulose (30–50%), hemicellulose (20–30%), and lignin (20–30%); with starch, protein, and oils as minor components.[11] The exact composition of each biomass varies depending both on the plant and on the residue collected.[1] Biorefinery can take advantage of the unique properties of each of biomass components, enabling the production of various products, so separation of these biomass components from each other seems meaningful. However, there are some difficulties in developing effective biomass separation processes because of the crystallinity of the cellulose, and the presence of covalent cross-linkages between lignin and hemicelluloses in the plant cell wall.[12] So, an efficient separation strategy includes: (1) disrupting and removing the cross-linked matrix of lignin and hemicelluloses that embeds the cellulose fibers, (2) disrupting hydrogen bonds in crystalline cellulose.[12]

Separation of biomass components is the very important initial step for the upcoming efficient utilization of biomass sources. Several physical and chemical separation methods are currently employed to overcome the recalcitrance of lignocelluloses. These methods include acid hydrolysis, alkaline hydrolysis, ammonia fiber expansion, hot water, organic solvent, and ionic liquid separation technologies.[2] In this section of the chapter, all of the above separation methods are discussed.

7.2.1   Acid Hydrolysis for Biomass Separation

Both concentrated and dilute acids such as H_2SO_4 and HCl have been used for biomass separation. Acid treatment solubilizes the cellulose and hemicelluloses and thereby disrupts the lignocellulosic composite material linked by covalent bonds, hydrogen bonds, and van der Waals forces. However, acids are toxic, corrosive, hazardous, and thus require reactors that are resistant to corrosion, which makes the separation process very expensive. In addition, the acid must be recovered after hydrolysis to make the process economically feasible.[13]

7.2.2   Alkaline Hydrolysis for Biomass Separation

Some bases can be used for the separation of biomass, and the effect of alkaline treatment depends on the lignin content of the materials.[3] Sodium, potassium, calcium, and ammonium hydroxides are suitable alkaline treatment agents. Alkaline treatment removes hemicelluloses and lignin and leaves cellulose. Compared with acid processes, alkaline processes cause less sugar degradation, and many of the caustic salts can be recovered and/or regenerated. The limitations and disadvantages of this kind of separation method are the long residence times required and the irrecoverable salts formed.

7.2.3   Ammonia Fiber Expansion for Biomass Separation

Ammonia fiber explosion (AFEX) is a physicochemical separation process in which lignocellulosic biomass is exposed to liquid ammonia at high temperature and pressure for a period of time, and then the pressure is suddenly reduced. AFEX treatment could remove lignin and hemicellulose to an extent. In a typical AFEX process, the dosage of liquid ammonia is 1–2 kg of ammonia per kg of dry biomass, the temperature is 90 °C, and the residence time is 30 min. A schematic apparatus for laboratory AFEX pre-treatment of biomass is shown in Figure 7.1. The disadvantage of AFEX is that it is not efficient for biomass with high lignin content.[14]

7.2.4   Hydrothermal Treatment for Biomass Separation

Passing hot water continuously through a stationary biomass bed could result in higher hemicellulose sugar recoveries, more lignin removal, more digestible cellulose, and less inhibitor in the hydrolysate liquid compared to conventional systems. The dissolved hemicelluloses are almost all in oligomeric form and always accompanied by lignin removal. Chemical reaction factors are not the only ones controlling hemicellulose hydrolysis and delignification reactions – mass transfer and other physical effects may play an important role in hemicellulose and lignin degradation.[15] Although the hydrothermal process does not require the acid resistant reactor materials of

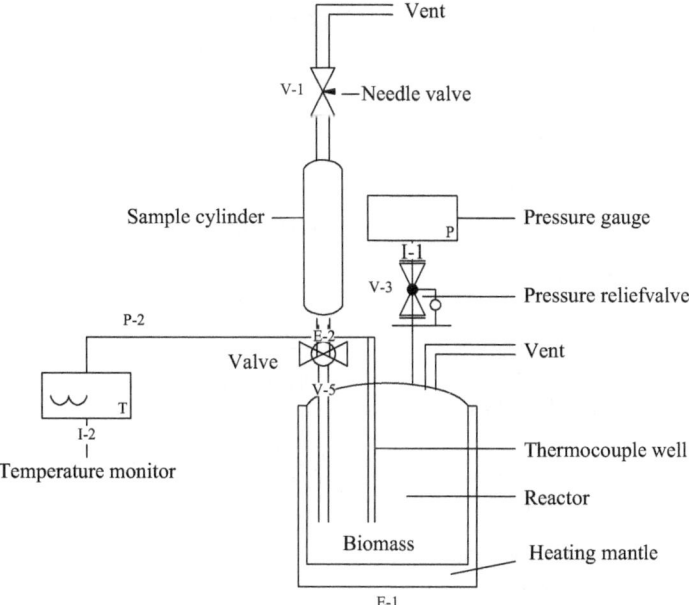

Figure 7.1 Schematic diagram of laboratory AFEX apparatus. (Reproduced from ref. 14.)

acid treatment, this advantage may be offset by increased water use and recovery costs.

7.2.5 Organic Solvent for Biomass Separation

Organic solvent separation has been shown as the separation process with the most potential that is currently under commercial development. For example, Lignol Innovations (Vancouver, Canada) has developed a bio-refining technology that employs an ethanol-based organosolv step to separate lignin, hemicellulose components, and extractives from the cellulosic fraction of woody biomass. However, the presence of strong hydrogen bonds between lignin and cellulose means that lignin micromolecules cannot be removed completely from the surface of cellulose, which limits the efficiency of lignin isolation.[16]

7.2.6 Ionic Liquid for Biomass Separation

The application of ionic liquid to biomass separation recently started to attract a great deal of attention. Cellulose is poorly soluble in conventional solvents due to its many intermolecular hydrogen bonds. Ionic liquid was believed to be capable of disrupting such hydrogen bonds between different polysaccharide chains, thus decreasing the compactness of cellulose and

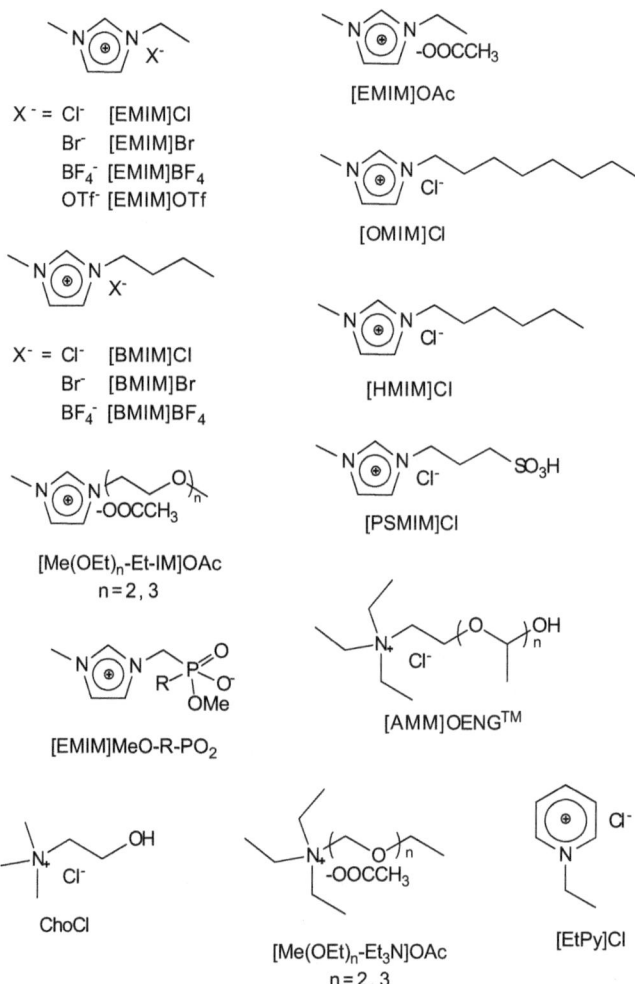

Figure 7.2   Examples of commonly used ionic liquid for cellulose dissolution and biomass separation.[17]

making the carbohydrate fraction more susceptible to hydrolysis. Examples of some of the ionic liquids used for cellulose dissolution and biomass separation are summarized in Figure 7.2.[17]

A novel method was developed recently to design 1-butyl-3-methylimidazolium bromine ([Bmin]Br) and aqueous ethanol solvent mixtures based on Hansen's theory of solubility, aiming at selective separation of wood.[18] The scheme of the treatment is shown in Figure 7.3. After treatment at 170 °C for 2 h, the cross-linked matrix between lignin and hemicellulose was destroyed, and the hydrogen bonds between cellulose and lignin were disrupted. Cellulose and lignin were obtained with purity of 94% and 93%.

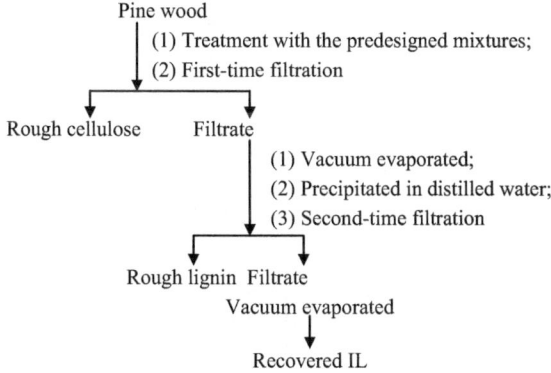

Figure 7.3 Scheme of the treatment of pine wood. (Reproduced from ref. 18.)

Recent studies have demonstrated that the combination of an acid catalyst and ionic liquid can be an efficient system for the hydrolysis of lignocellulosic materials under mild conditions with improved yields of HMF or total reducing sugars. Reducing sugars are compounds possessing one aldehyde or ketone group which allows the sugar to act as a reducing agent. Researchers have also hydrolyzed corn stalk, rice straw, pine wood, and bagasse in 1-butyl-3-methylimidazolium chloride in the presence of hydrogen chloride at 100 °C within 60 min.[19] The total reducing sugars yields could reach up to 81%. Another report showed that the acid-catalyzed conversion of loblolly pine could be conducted at mild temperature (120 °C) in 1-butyl-3-methyl imidazolium chloride; and the carbohydrate fraction could be completely converted into water-soluble products (monosaccharides, oligosaccharides, furfural, and HMF).[20] In acid catalysis conditions, the lignin remained as a solid residue, so complete sugar–lignin separation was achieved.

Transition metals have been reported to be good catalysts for the conversion of sugars to useful intermediates in ionic liquid. The combination of $CrCl_2$ and an ionic liquid system has been recently reported in effectively catalytic conversion of glucose to HMF with 70% yield under mild conditions. The dehydration of fructose to HMF using metal halide catalysts (including $CrCl_2$, $CrCl_3$, $FeCl_2$, $CuCl$, $CuCl_2$, or $RuCl_3$) at low temperature (80 °C) has been reported. HMF yields ranging from 63% to 83% were obtained when these metal halide catalysts were added into 1-ethyl-3-methyl-imidazolium chloride. Moreover, these reactions had good selectivity and only a negligible amount of by-products were formed.[21]

Extraction of lignin from bagasse using the ionic liquid mixture [C2mim][ABS] was successfully achieved at atmospheric pressure with over 93% yield[22] and the extracted lignin was of a relatively uniform molecular weight. In their study, the compositions of ionic liquid mixture

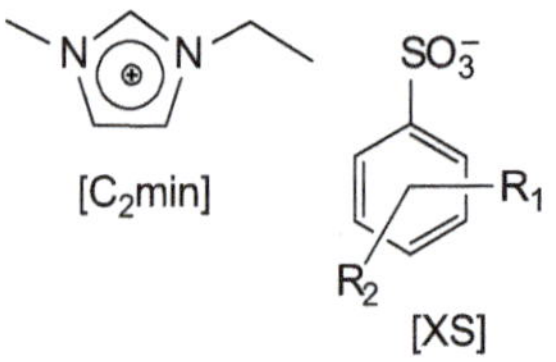

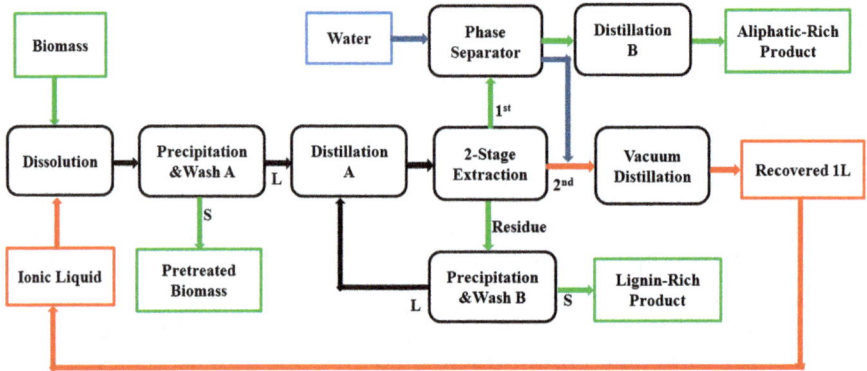

Figure 7.4   Ionic liquid 1-ethyl-3-methylimidazolium xylenesulfonate.[22]

Figure 7.5   Biomass separation system and the recovery of used ionic liquid.[23]

were 1-ethyl-3-methylimidazolium [C2mim] (a cation) and alkylbenzene-sulfonate [ABS] (a mixture of anions). The anions were mainly made up of isomers of xylenesulfonate (74%) with smaller amounts of ethylbenzene-sulfonate (13%), cumenesulfonate (9%), and toluenesulfonate (4%). The main component of [C2mim][ABS] is shown in Figure 7.4. One advantage is that the process can be conducted at atmospheric pressure, which is significant in terms of plant and equipment requirements.

However, ionic liquid is expensive and it should be used effectively. So after reactions, efficient recovery and removal of residual biomass solutes from ionic liquid is an important step, which is necessary for recycling the ionic liquid and in an ionic liquid separation process.

The recovery of the ionic liquid 1-ethyl-3-methylimidazolium acetate (abbreviated as [C2mim][OAc]) with a solvent mixture containing acetone, 2-propanol, and a small amount of water was demonstrated.[23] This formed a phase switchable solvent system, which could concentrate and separate oleophilic solutes, short chain carbohydrates, and lignin fragments from used ionic liquid in a two-stage extraction procedure. Figure 7.5 shows the whole process of ionic liquid recovery and biomass separation. The research was conducted by pre-treating 100 g of corn stover with 10 wt.% of ionic liquid. After the separation process, the recovered ionic liquid contained few residual solutes, so it could be reused for biomass dissolution and separation. The above work is meaningful, because this process does not intrinsically generate chemical waste, and it complies with the principle of green chemistry by reusing ionic liquid solvent.

7.2.7 Biomass Separation and Production of Chemicals and Fuels

Varieties of methods for lignocellulosic biomass conversion have been studied. Catalytic hydrolysis, solvolysis, liquefaction, pyrolysis, gasification, hydrogenolysis, and hydrogenation are the major processes presently studied.[24] Liquefaction and fast pyrolysis of biomass are primarily conducted at relatively high temperature. Gasification is typically conducted over supported noble metal catalysts. Most processes yield a complex mixture, leading to problematic upgrading and separation. Some techniques are to integrate hydrolysis, liquefaction, or pyrolysis with hydrogenation over multifunctional solid catalysts to convert lignocellulosic biomass into value-added chemicals and bio-hydrocarbon fuels.

Alonso *et al.*[25] have eliminated pre-treatment steps to fractionate biomass. They made use of certain composition in biomass to obtain fuels and chemicals, and the rest of fraction was separated. The work used gamma-valerolactone (GVL) as solvent, and the cellulosic fraction of lignocellulosic biomass can be converted into levulinic acid (LA), while at the same conditions the hemicellulose fraction can be converted into furfural. The furfural can be separated by distillation during the reaction or can be kept in the reactor and subsequently processed to produce furfuryl alcohol and LA. The lignin was solubilized in the GVL and separated. This process not only obtains the production of fuels and chemicals by utilization of hemicellulose and cellulose, but also it benefits from the elimination of pre-treatment and extraction/separation steps.

Bio-oil, a kind of biofuel, could be obtained by flash pyrolysis of biomass. Because bio-oil is a type of complex mixture, some researches about bio-oil separation have been conducted in the recent years. Xianwei Zheng *et al.* have successfully separated flash pyrolysis oil into four kinds of substances by united extraction and distillation.[26] Garcia-Perez *et al.* have provided an efficient separation method by using five kinds of solvent extraction, and bio-oil was separated into six kinds of substances.[27]

7.2.8 Summary

Lignocellulosic biomass is the most abundant and bio-renewable resource with great potential for sustainable production of chemicals and fuels. The conversion of biomass to biofuels begins with biomass pre-treatment to remove or weaken the tight linkages among cell wall components, making biomass easier to separate. The process of biomass separation is one of the most difficult and expensive operations. So an economical and effective separation or pre-treatment method is a great challenge for researchers.

7.3 Chemicals Production From Lignin

Lignin, which accounts for 15 to 30 wt% of woody biomass, is also available from agricultural residues. Lignin is a cross-linked amorphous copolymer synthesized from random polymerization of three primary phenylpropane monomers which are bonded together through several different C–O–C and C–C inter-unit linkages, and the dominant bond is the β-O-4 linkage. The role of lignin in biomass is to provide the plant with structural integrity, water impermeability, and resistance against microbial decay.[28] The future economic viability of lignin biorefineries depends on conversion of lignin to value-added products.

Although lignin holds great potential as a renewable source of fuels and aromatic chemicals, lignin utilization technologies are substantially less developed. Difficulties in catalytic processing caused by the existence of a variety of different inter-unit linkages, high affinity for the formation of a more condensed structure when thermochemically processed, poor product selectivity, and ease of use as a solid fuel are the major barriers towards the development of a lignin-based biorefining technologies. Owing to the massive amounts of lignin available in the pulp mills and in future biorefineries, establishment of lignin conversion processes will open selectable routes for the production of low carbon biofuels and chemicals.[29]

7.3.1 Lignin-based Materials

Lignin is part of the composition of natural polymers in variable proportions. The aromatic structure of the lignin can be used as source of several phenolic products, which may substitute petroleum-based compounds. Bio-based composites have gained prominence over the past two decades owing to both environmental concerns and waste disposal problems. Lignin-based biomaterials include carbon fibers, polymer modifiers, resin/adhesives/binders and others.

7.3.2 Lignin Application in Carbon Fibers

As important engineering materials in advanced composites for a variety of industries, carbon fibers were applied to aerospace, civil engineering, automotive, and wind-power applications. Their properties were characterized by high tensile strength, high stiffness, low density, elevated temperature tolerance, and low thermal expansion.[30] Within a considerable amount of research in the past decade, lignin was proved to be an appropriate precursor to produce carbon fibers.[31] And as one of three major biopolymers in the cell wall of plants, with the second most abundant terrestrial material, lignin is considered to be a very suitable, readily available, relatively

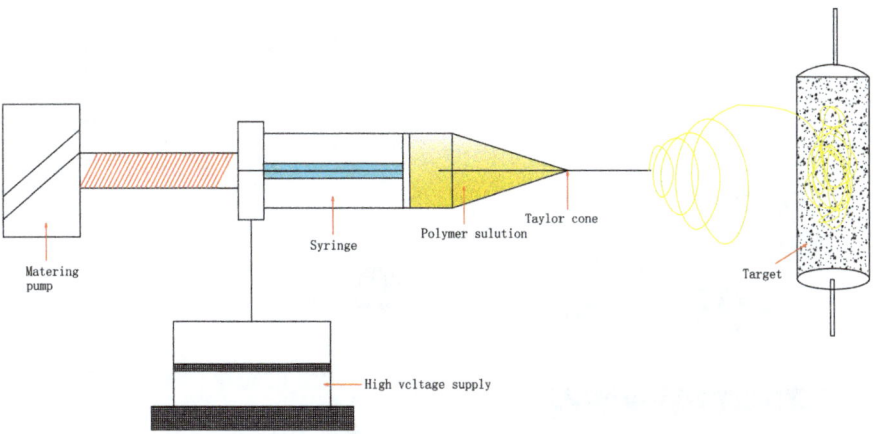

Figure 7.6 Lignin-based carbon fibers prepared by electrospinning.

inexpensive, and renewable carbon fiber precursor. Today, 90% of the raw material for manufacturing carbon fibers is polyacrylonitrile (PAN), coal- and petroleum-derived pitches, and regenerated cellulose (Rayon), while the areas of application of CFs are limited because of the high production cost; lignin is a possible material with low costs that has increased the interest in identifying alternative fiber precursors. Moreover, lignin-derived CFs should be inherently both low and insensitive to changes in petroleum price.

Several studies investigated the possibility of using various technical lignins (*e.g.*, lignosulfonates, organosolv lignins, steam explosion lignins, and kraft lignins) as raw material for carbon fiber production. The kraft lignin-based carbon fibers with the best mechanical properties were obtained from hardwood kraft lignin with 5% polyethylene terephthalate (PET) added. With the spinning (as shown in Figure 7.6), the spinnability of hardwood kraft lignins is often more satisfactory than softwood kraft lignin, but stabilization is a time-consuming step. Stabilization aims to transform a spun fiber to a thermoset character, detailed methods including crosslinking, oxidation, and cyclization reactions. During thermal treatment of lignins, the actual temperature was maintained lower than the glass-transition temperature (T_g), model calculations performed predicted that a temperature increase of less than 0.06 °C min^{-1} is required to hinder the fusion of the fibers. Furthermore, hardwood kraft lignins purified using organic solvents required an even lower heating rate, as low as 0.05 °C min^{-1}, to maintain intact fibers after carbonization.

7.3.3 Polymer Modifiers

As a renewable resource, lignin is generated by plants *via* photosynthesis using carbon dioxide from the atmosphere then followed by aromatization

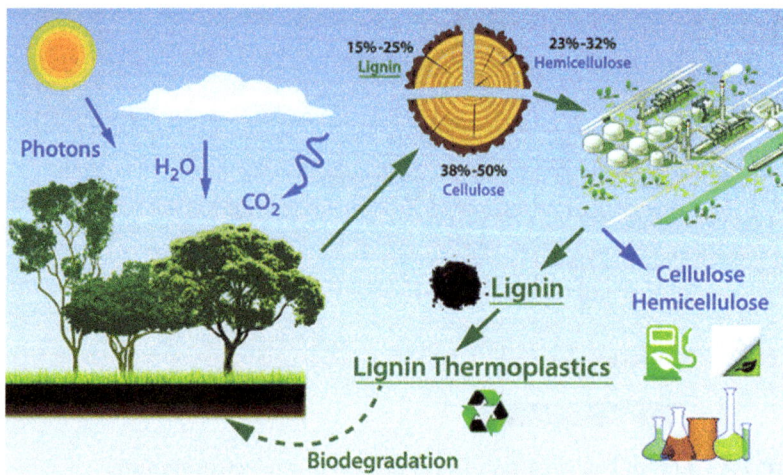

Figure 7.7 Production of natural lignin and lignin-derived polymer and their recycling.[33]

and polymerization of carbohydrates (Figure 7.7). Since it has good compatibility with polymers, lignin can be incorporated into thermoplastic polymers (*e.g.*, polyolefins, polyacrylics, PVC) as a polymer modifier.[32] It is a potential start for lignin to be a low cost additive, and its use in polyurethane formulations could displace petroleum-derived compounds, which can also improve the thermal and mechanical properties of kraft lignin urethanes. However, in order to avoid any environmental problems caused by the release of sulfur-containing gases, sulfur-free lignins such as organosolv or soda lignin should be used.

As we know that phenol-formaldehyde (PF) resins were the first completely synthetic commercial polymer. Lignin constituted by structure of three primary phenylpropane monomers can replace phenolic compounds in the synthesis of phenol-formaldehyde (PF) resins. Besides its use in polyurethanes and polyesters, technical lignins are also of interest in phenolic and epoxy resins. Kraft lignin can be used to displace up to 70% of the phenol required for PF resins, which can greatly decrease the raw material consumption.

The first prototype printed circuit board for the electronics industry was formulated by IBM with a 50% lignin-containing epoxy resin. Kraft lignins are insoluble in water, which can make advanced modifications products such as asphalt emulsifiers. The wet strength of kraft liner has been reported to increase by laccase-aided grafting of lignin model compounds. An increased wet strength in kraft liner could therefore be facilitated by use of black liquor lignin derivatives with high free phenolic groups. The use of kraft lignin derivatives in this application could become a large scale business as 25–30% of the world paper production.

7.3.4 Resin, Adhesives, Binders and Others

Due to the presence of sulfonic groups, characteristics of the lignosulfonate are quite different from those of kraft lignins. Compared with other lignins, lignosulfonates expand their applications as a result of their unique colloidal properties. As we know, lignosulfonates can be used as dispersing agents for oil well drilling products, detergents, stabilizers, surfactants, adhesives, cement additives, battery expanders, animal feeds, raw materials for the preparation of dyestuffs and pesticides, *etc.*

Their usage as cement additives was one of the first large products based on lignosulfonates. Water-reducing concrete admixture is a surface-active chemical which can disperse and improve the fluidity and workability of fresh concrete. As we know, the less water used the higher the strength of the concrete after solidification. For the same strength requirements, the amount and weight of the concrete will be reduced. Lignosulfonates have been extensively studied for their adhesive properties with poor reproducibility of the bonding effects, due to variable properties of lignin from various sources.

In dye industry, lignin dye dispersants were favored in reducing the rate of coloring, stabilizing dispersion at high temperatures (above 100 °C) and inertia to the dye molecules. Digestibility of dietary fiber and crude protein in animals was inhibited, with cellulose and lignin being the major determinants for changes in digestibility. Lignin was found to have a strong negative influence on fiber digestion and was undigested in both the small and large bowel of humans. Lignin extracts from corn stover residues generated during ethanol production were shown to exhibit antimicrobial activities against pathogenic bacteria and yeast, which could be an application in antimicrobial packaging. Research is underway to demonstrate the use of lignin nanotubes as carriers of cancer-fighting drugs. Organosolv lignins with more phenolic and fewer aliphatic hydroxyl groups with low molecular weight and narrow polydispersity were reported to have high antioxidant activity.

7.3.5 Power, Fuels and Syngas Products

Depending on the process conditions, combustion, gasification, pyrolysis, and liquefaction can degrade lignin to a different extent. From partial depolymerization to low molecular weight lignin fractions, lignin can be converted to power, liquid fuel, and syngas products.[34]

Contributing as much as 40% of the energy content of lignocellulosic biomass, lignin has a heating value of nearly 18 MJ kg^{-1}.[35] Combustion of lignin from black liquor to produce process heat, power, steam and recovering pulping chemicals is widely practiced today.

Lignin (black liquor) gasification (Figure 7.8) converts it to a gaseous fuel (syngas) when carried out at high temperature with or without partial oxidation. In addition to syngas, the gas stream contains water vapor, carbon

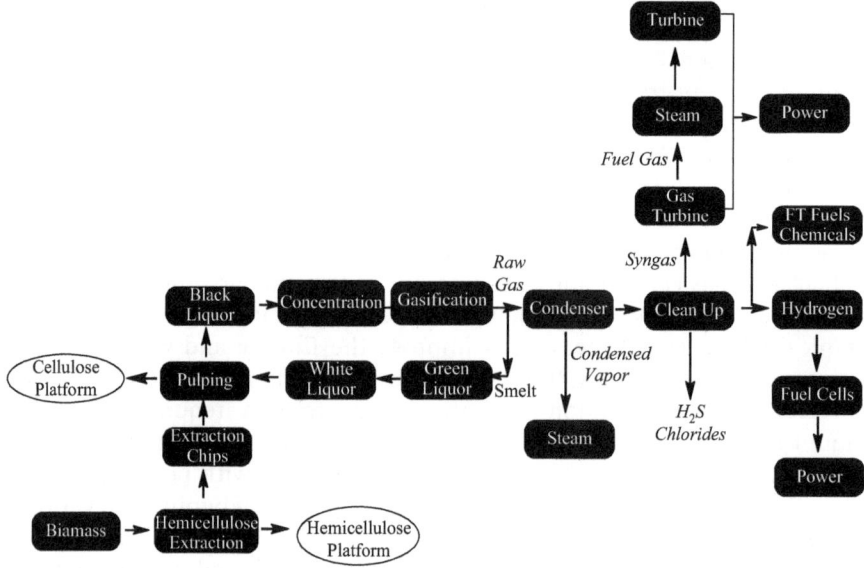

Figure 7.8   Black liquor gasification.[37]

dioxide, nitrogen, ammonium, hydrogen sulfide, hydrogen chloride, methane, *etc.* Syngas can not only be used for heat and power applications, but can also be used for production of hydrogen, methanol, ethanol, and hydrocarbons (wax, diesel, gasoline, and naphtha). However, there are still problems in the gasification technology, including deactivation of catalyst, greenhouse gas emissions, corrosion, water and air pollution, clogging of equipment, *etc.* It is expected that costs will decrease with the development of gasification technology and increasing scale of production.[36]

Fast pyrolysis of lignin produces three kinds of fractions: liquids, gases (hydrogen, methane, carbon dioxide, carbon monoxide, *etc*), and solids (bio-char). Typically, fast pyrolysis (673–973 K, hold for 0.3–15 s) results in 30–50% liquid, 30–50% bio-char, and 6–40% gases.[38] Rapid heating of the pyrolysis reactor and cooling of pyrolysis vapors are two important parameters that can maximize the liquid yield. Bio-char can be utilized as solid fuel, for production of activated carbon and other chemicals, for carbon sequestration, bioremediation, and for improving soil functions such as soil erosion, and water and nutrient retention. Liquid products contain a high molecular weight fraction (pyrolytic lignin), monomeric phenolic compounds, and low degradation compounds. Monomeric phenolic compounds are typically characterized by dimeric and monomeric biphenyl, phenyl coumaran, diphenyl ethers, stilbene, and resinol structures. Low degradation compounds include methanol, hydroxyacetaldehyde and acetic acid. Therefore, pyrolysis may be useful as a technology for the controlled molecular weight reduction of lignin that can offer some unique possibilities for conversion to useful aromatics. The lignin-derived phenolic compounds can be recovered from the liquid and used to displace phenol in the production of PF resins. The major obstacle in the extraction of valuable products from bio-oil is their low concentration that currently renders recovery technically difficult and economically unattractive.

7.3.6 Liquefaction to Prepare Value Added Phenolic Compounds

Compared with lignin fast pyrolysis, liquefaction (pressure 10–20 MPa, 473–723 K) results in higher liquid yields (60%). As a replacement method for fast pyrolysis, liquefaction processes not only improve the small molecules in liquid production, but also dispense with the process of lignin purification. Lignin depolymerization products include phenol, cresol, guaiacol, eugenol, catechins, vanilla, benzene, toluene, naphthalene, *etc.*[39] However, the only industrial use of lignin was in the production of vanillin, which today is mainly based on nitrobenzene oxidation of lignosulfonates under alkaline conditions.

Vanillin is a phenolic aldehyde used in the food industry as a flavoring agent, mainly applied in ice cream and chocolate industries, with smaller amounts being used in confectionary and baked goods. Vanillin and related phenols can also be produced by microbial degradation of lignin.

Lignin can be selectively used to produce 4-ethylphenolics *via* mild hydrogenolysis (Figure 7.9). 3.1% 4-ethylphenol and 1.3% 4-ethylguaiacol could be obtained simultaneously from hydrogenolysis of cornstalk lignin, which is approximately the yield obtained from the petrochemical route. The result indicates that hydrogenolysis of lignin is a quite promising technique for the substitution of petrochemical route to get 4-ethylphenol and 4-ethylguaiacol.

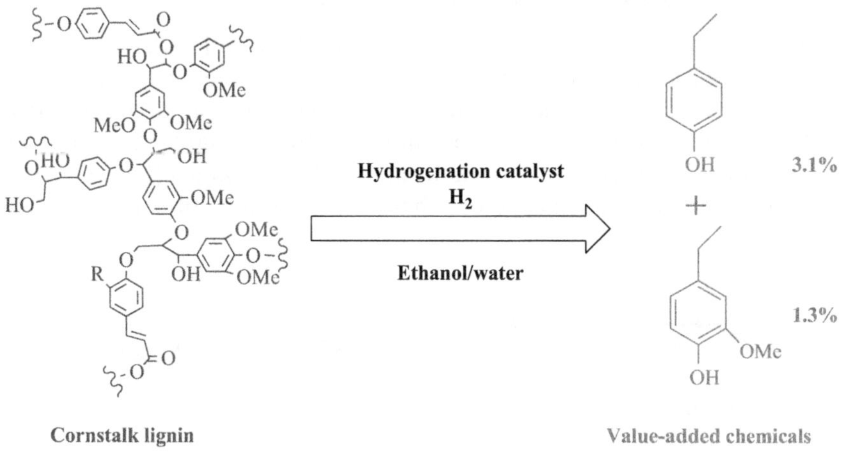

Figure 7.9 Cornstalk lignin is selective to produce 4-ethylphenolics.[40]

7.3.7 Summary

Currently, lignin is being utilized as a low grade boiler fuel, but several studies have been done to convert lignin to value-added products. These efforts are beneficial for biofuel production, whereas less attention has been paid to the separation of lignin. Main pathways for the utilization of lignin are:

1. separation and purification of lignin from current lignin products;
2. exploiting an efficient biorefinery approach to convert lignin to value-added chemicals.

7.4 Upgrading of Cellulose and its Products

Cellulose, a polydisperse linear 1,4-β-glucan, is part of a renewable resource which is the most abundant natural polymer on earth. It has been estimated that the global production of cellulose is 1.5 trillion tons each year, and is considered an almost inexhaustible source of raw material for environmentally friendly and biocompatible products.[41] Cellulose is widely used for coatings, membranes, pharmaceuticals, and foodstuffs.[42]

Nowadays, research on cellulose for chemicals and polymeric materials are mainly focused on the following aspects:

1. the catalytic conversion of cellulose to fuels and chemicals;
2. the development of environmentally new solvent systems to dissolve cellulose and following applications;
3. the modification of functional materials with cellulose derivatives or cellulose graft copolymers.[43]

7.4.1 Structures and Properties of Cellulose

A cellulose molecule has the generic chemical formula $(C_6H_{12}O_5)_n$. The structure and properties of cellulose have been the subject of a large amount of work. It consists of a skeletal linear polysaccharide, and connected by β-1,4-glycosidic linkages. The glucose units are further tightly bound by numerous extensive inter- and intramolecular hydrogen bonds (Figure 7.10).[44] The glucopyranose units of cellulose chains range from approximately 100 to

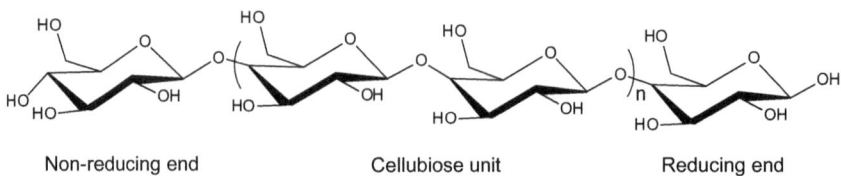

Non-reducing end   Cellubiose unit   Reducing end

Figure 7.10   The central part of a cellulose molecular chain.
(Reproduced from ref. 44.)

14 000. Accordingly, cellulose has an average molecular weight in the range of 300 000–500 000. One of the most interesting characteristics is that cellulose consists of several crystal polymorphs, with the possibility of conversion from one form to another. Its six different polymorphs differ in unit cell dimensions and chain polarity, and are the principle component of all plant cell walls. The natural cellulose I has two different structures, Iα and Iβ, while cellulose II is another important crystalline form of cellulose. The transformation of cellulose I to cellulose II is generally considered to be irreversible, because cellulose II is more stable than cellulose I. With proper chemical treatments, it is possible to produce cellulose III and cellulose IV.

7.4.2 The Catalytic Conversion of Cellulose to Fuels and Chemicals

Over decades, many researchers have attempted to explore the production of fuels and chemicals from cellulose. Figure 7.11[45] lists some typical fine chemicals and fuels which can be produced from cellulose.

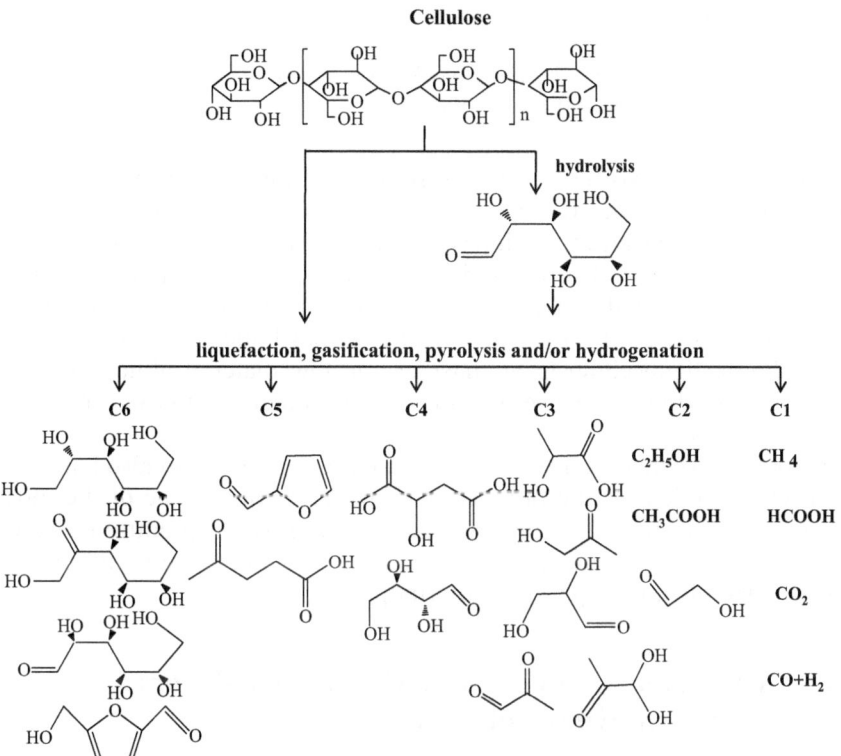

Figure 7.11 Potential chemicals and fuels from the catalytic conversion of cellulose. (Reproduced from ref. 45.)

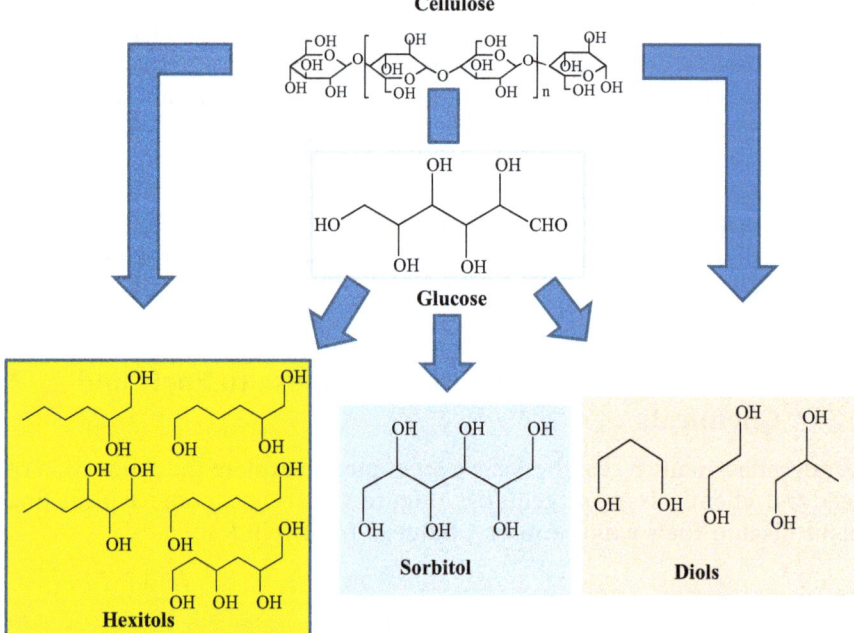

Figure 7.12 Production of building blocks from glucose dehydration.
(Reproduced from ref. 46.)

Cellulose hydrolysis has received serious attention because many attract-ive building blocks and fine chemicals can be prepared from further con-version of reducing sugars. Cellulose and glucose also can be converted into several polyalcohols by hydrogenation. The main products obtained by these reactions are depicted in Figure 7.12.[46] Generally, taking into account the cost of pyrolysis, gasification, and hydrothermal liquefaction, researchers often choose lignocellulosic biomass as the raw material although cellu-lose can be directly converted into value-added chemicals in these processes alone.

Biological enzymatic conversion of glucose by fermentation also has been extensively studied for the production of many products; one of the most important is bioethanol. Many bacteria and fungi could produce enzymes for the hydrolysis of cellulosic material. These microorganisms can be aer-obic or anaerobic, mesophillic or thermophillic.[47]

7.4.3 Functional Cellulose Materials Prepared from Non-derivatizing Solvents

Cellulose is very difficult to dissolve in common solvents due to its intra-molecular and intermolecular hydrogen bonding networks. Lately, many new non-derivatizing solvents have been developed to dissolve cellulose, such as *N*-methylmorpholine-*N*-oxide, ionic liquids (ILs), LiCl/dimethylace-tamide system (LiCl/DMAc) and alkali/urea (or thiourea) aqueous systems.[48] They provide opportunities for the preparation of functional materials, in-cluding cellulose-based hydrogels, electrospun fibers, and cellulose beads, which have many advantages including being safe, biocompatible, and biodegradable. In recent years, our laboratory has focused on dissolving cellulose in ILs and the following applications.

7.4.3.1 Hydrogels

The development of LiCl/DMAc as a solvent for cellulose has promoted the research of hydrogel production from native cellulose. Cellulose materials with a large molecular weight range can be dissolved in the LiCl/DMAc so-lution. Although the use of *N*-methylmorpholine-*N*-oxide provides a simple method to produce regenerated cellulose fibers, films, and beads, rare transparent hydrogels are prepared directly from the cellulose solution in the *N*-methylmorpholine-*N*-oxide system. Cellulose hydrogels have been obtained by dissolving in 1-allyl-3-methylimidazolium chloride, then followed by regeneration using liquid nitrogen freeze drying to prepare nanoporous cellulose foams successfully.[49] A novel alkali/urea or thiourea aqueous system was developed, in which cellulose ($M_w < 1.2 \times 10^5$) could be dissolved rapidly; thereby, cellulose hydrogels can be obtained *via* physical cross-linking of the macromolecular chains by increasing temperature.[48]

7.4.3.2 Electrospun Fibers

Electrospinning is a method of forming fibers through the action of elec-trostatic forces. Several excellent reviews in electrospinning have been published to introduce its merits and applications. The LiCl/DMAc solvent for cellulose is a solution of a non-volatile salt in a volatile solvent. Diameters of fibers depend significantly on cellulose DP; for example, fibers spun from high molecular weight (DP > 1000) cellulose had very fine intermittent segments.[50] Recently, cellulose was directly spun from *N*-methylmorpholine-*N*-oxide/water commercially *via* dry-jet wet spinning.[49] Cellulose-based fibers were prepared by electrospinning from cellulose dissolved in alkali/urea aqueous systems in the presence of polyol binders by Qi *et al.*[51] Cellulose fibers also have been prepared by electrospinning cellulose from 1-butyl-3-methylimidazolium chloride solution into an ethanol bath.[48]

7.4.3.3 Cellulose Beads

Cellulose beads are spherical particles with diameters in the micrometer to millimeter scale, which are used in chromatography, solid-supported syn-thesis, protein immobilization, drug release, and so forth. In principle, bead production can be simplified into two procedures: dropping and dis-persion. Gericke[52] has reported a method of preparing beads by dropping cellulose/DMA/LiCl solutions into coagulation media and a novel approach to obtain cellulose beads by the formation of spherical cellulose/NMMO droplets, either by dropping or dispersion, at temperatures around 20–40 °C. Cellulose composite beads also have been prepared successfully from alkali/urea (or thiourea) aqueous systems by dropping techniques for removal of heavy metal ions, including spinning drop atomization. Certain ionic liquids (ILs) already have been successfully employed for the preparation of cellulose beads *via* dispersion and dropping techniques.[53]

7.4.4 Derivative Modification of Cellulose and Graft Copolymers

Cellulose is a unique biopolymer and possesses several merits, and thus is a primary candidate for biological materials. Unfortunately, cellulose has some inherent shortcomings, such as poor solubility in common solvents, unsatisfactory mechanical properties, and poor dimensional stability. However, one of the appropriate approach to solving this problem is to bring functional groups into cellulose molecules through chemical modification, which could introduce new properties to the cellulose.

At present, important commercial cellulose esters include cellulose acetate butyrate, cellulose acetate propionate, and cellulose acetate. In recent developments, a variety of cellulose chemicals, such as quaternized cellulose, carboxymethyl cellulose, methyl cellulose, 3-allyloxy-2-hydroxypropyl cellulose, and the hydroxyalkyl celluloses, have been synthesized *via* etherification reactions. In addition, cellulose derivatives have many applications, including antioxidant agents, shape memory gels, photoactive materials, gene carriers, and in memory devices.[43]

Graft copolymerization offers a promising method to impart functional groups to a polymer among all the methods of modification of polymers. In 1943, the pioneering work of Ushakov obtained some insoluble products that were probably the first graft copolymers. Since then, a variety of studies have been begun to focus on the synthesis and applications of cellulose graft copolymers.[54] Generally, approaches to synthesis of cellulose graft copolymers include controlled radical polymerization, ionic and ring-opening polymerization, and free radical polymerization. Figure 7.13 illustrates a schematic structure of a cellulose graft copolymer.[54]

7.4.5 Summary

Cellulose has been considered as one of the most attractive alternatives to replace fossil resources and it plays an important role in the functions of plants. New findings in organic chemistry and biology should receive more attention in future research. Cellulose is also the most abundant renewable natural polymer and is important in both polymer science and materials science. Cellulose derivatives and graft copolymers are promising as blocks

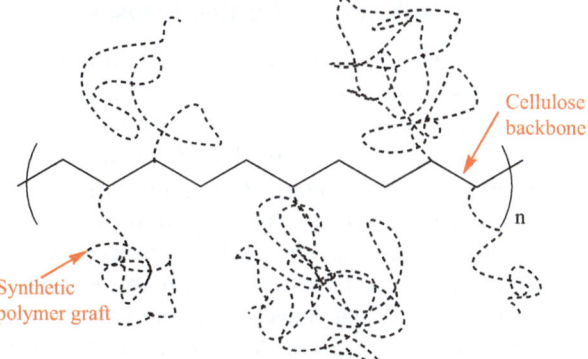

Figure 7.13 A schematic representation of cellulose graft copolymer. (Reproduced from ref. 54.)

for the fabrication of functional materials. Efforts should be made to find versatile approaches for the preparation of cellulose graft copolymers with relevant architectures of both graft chains and cellulosic backbones.

7.5 Chemicals and Materials from Hemicellulose

Recently, hemicelluloses have gained increasing importance as a basis for new biopolymeric materials and functional polymers accessible by chemical modification. Hemicelluloses belong to the biopolymers that are non--crystalline heteropolysaccharides and are classically defined as the alkali-soluble material remaining after the removal of the pectic substances. Hemicellulose is the second most abundant constituent of lignocellulosic biomass after cellulose. The growing willingness to develop new biopolymer-based materials has led to an increasing application of hemicelluloses and their derivatives. Hemicelluloses can be converted into a variety of low molecular mass chemicals such as furfural, ethanol, and xylitol. Hemicelluloses are also attractive as biopolymers, which can be utilized in their native or modified forms in various areas.[55]

7.5.1 Fractionation and Purification of Hemicelluloses

Hemicelluloses are non-cellulosic and short-branched chain hetero-polysaccharides, which consist of various different sugar units. They can be arranged in different proportions and with different substituents. Large amounts of hemicelluloses with a wide variation in content and chemical structure are found in plant cell walls. Hemicelluloses generally consist of several populations of polysaccharide molecules, varying in structural characteristics. Several fractionation techniques have been employed in order to obtain more homogeneous fractions as well as exploring structure–property relationships for the hemicellulosic polymers.[56]

7.5.2   Enzymatic Hydrolysis of Hemicelluloses

As we know, hemicelluloses can be hydrolyzed into mono and oligo-saccharides by enzymes. Compared to monosaccharides, the functional oligosaccharides present important physicochemical and physiological properties, which are beneficial to the health of consumers, and they have been extensively used as pharmacological supplements and food ingredients, reducing serum lipid levels in hyperlipidemics.[57] In recent years, among the functional oligosaccharides, xylo-oligosaccharides (XOS) have received special attention due to their favorable features including stability in acidic media, resistance to heat, lower available energy, and significant biological effects at low daily intakes. This section is mainly about the current research surrounding the manipulation of XOS by enzymes.[58] The current knowledge about the enzymatic hydrolysis of hemicelluloses for XOS production is summarized in Table 7.1.

7.5.3   Materials and Chemicals from Hemicellulose

7.5.3.1   *Biofuels and Chemicals from Hemicellulosic Fermentation*

Processes based on microbial fermentation are currently regarded as alternatives having the most potential in converting hemicelluloses into biofuels and chemicals such as ethanol, butanol, hydrogen, and succinic acid due to the advantages of low cost and environmental friendliness.[59] Recent studies on biofuels and chemicals from hemicellulosic fermentation are shown in Table 7.2.

7.5.3.2   *Production of Fuel Ethanol from Hemicellulosic Hydrolysates*

In utilizing hemicellulosic sugars, efficient and cost-effective conversion of lignocellulose materials to fuel ethanol is essential. For production of fuel ethanol, various waste and under-utilized lignocellulosic agricultural residues can serve as low-cost feedstocks.[60] Any hemicellulose that contains lignocellulose generates a mixture of sugars upon pre-treatment alone or in combination with enzymatic hydrolysis.[61] Xylose can be converted to xylulose *via* the enzyme xylose isomerase, while traditional yeasts can ferment xylulose to ethanol. By using two different approaches, several microorganisms have been genetically engineered to overproduce ethanol from mixed sugar substrates. The first approach is to divert the carbon flow from native fermentation products to ethanol in efficient mixed sugar utilizers such as *Escherichia*, *Erwinia*, and *Klebsiella*, and the second approach is to introduce the pentose-utilizing capability into efficient ethanol producers such as *Saccharomyces* and *Zymomobilis*.[62]

Competitive and economical production of fuel ethanol from hemicellulosic biomass holds strong promise. Much global research effort is

Table 7.1 Enzymatic hydrolysis of hemicelluloses for XOS production.[56]

Substrate	Origin	Enzyme	Main results
Arabinoxylans (5 g L^{-1})	Rye	Endo-1,4-β-D-Xylanase (10,000 nkat xylanase/g AX)	α-L-Araf-(1→3)-β-D-Xylp-(1→4)-D-Xylp
Xylan (2%)	Cotton stalks	Commercial xylanase (1.1 units mL^{-1})	XOS in the DP range of 2–7 (X6≈X5>X2>X3) with minor quantities of xylose
Xylan (1%, w/v)	Birchwood and wheat bran	ReBlxA (122.9 U mg^{-1})	Xylotriose (X3)
Xylan (2%)	Tobacco stalk, cotton stalk, Sunflower stalk and wheat straw	Xylanase from *A. niger* (4 U mL^{-1}); Xylanase from *T. longibrachiatum* (4 U mL^{-1})	X2>X3>X4>X5>X6, >X6
(Glucurono)arabinoxylans (1%)	Barley husks	Endo-β-Xylanase (5 U mL^{-1})	β-D-Xylp-(1→4)-[α-L-Araf-(1→3)]-β-D-Xylp-(1→4)-β-D-Xylp-(1→4)-β-D-Xylp
Arabinoxylan (1.0%, w/w)	Wheat	Combination of Ultraflo L (5%, w/w), β-xylosidase (0.25 g kg^{-1}), esterase (0.1 g kg^{-1}) and acetyl xylan esterase (0.1 g kg^{-1})	Ferulic acid (1.6 mg), acetic acid (24 mg), arabinose (51 mg), xylose (167 mg), and solubilized oligosaccharides (244 mg) per gram substrate dry matter
Water-unextractable arabinoxylans (1 wt.%,)	Birchwood and wheat bran	*B. subtilis* xylanase (0.12 μmol g^{-1})	Molecules of high (DP>25), medium (DP 9–25), and low (DPb9) molecular weights
Water extractable polysaccharides (50 mg mL^{-1})	Wheat bran and bengal gram husk	Driselase (0.28 U mg^{-1})	→4][α-L-Araf-(1→2)][α-L-Araf-(1→3)]-β-D-Xylp-(1→ Xylp-(1→

Table 7.1 (Continued)

Substrate	Origin	Enzyme	Main results
Xylan (4 mg mL^{-1})	Eucalyptus wood, brewery's spent grain	Endo-(1,4)-β-D-xylanase I (0.2 µg mL^{-1})	$X_nAc_m, X_n(GlcA_{me})_1Ac_m, X_n(GlcA_{me})2Ac_m$, and $X_n(GlcA_{me})_1$ or $_2Ac_mH$
Arabinoxylan (5 g L^{-1})	Oat spelts	Termamyl®120 L (0.7 U mg^{-1})	β-D-Xylp-(1→2)-a-LAraf(1→3)-β-D-Xylp-(1→4)-D-Xyl
Xyloglucan (0.1%, w/v)	Benincasa hispida	Endo-(1→4)-β-D-Xylanase (10 units)	XXXG type, and containing XXXG, XXFG, XXLG and XFLG as major oligomeric building sub-units
Xylan (0.1%, w/v)		Endo-(1→4)-β-D-Xylanase (22.5 units)	Exhibiting a classical structure with a backbone of β-(1→4)-linked xylopyranosyl residues substituted with three 4-O-methyl glucuronic acid per 97 xylopyranosyl unit
Insoluble dietary fiber (40 g L^{-1})	Wheat bran	Sunzymes (0.4%, w/w)	Degree of polymerization of 2–7 and the ratio of arabinose to xylose of 0.27, and XOS was strongly resistant to lower acidic conditions
4-O-Methyl-D-glucuronoxylan	Beechwood	Family 10 EXs, (0.2 U mL^{-1})	MeGlcA^{3}Xyl$_3$, MeGlcA^{3}Xyl$_5$
4-O-Methyl-D-glucuronoxylan (2%, w/v)	Birchwood	Endoxylanase (0.1 U mL^{-1})	Aldotetrauronic acid

Table 7.2 Biofuels and chemicals from hemicellulosic fermentation.[56]

Raw material	Type of pretreatment	Amount of solids	Strain	Cultivation on hydrolyzate	Product	Mode of operation
Salix chips	Steam of SO_2	9%, w/w WIS	Saccharomyces cerevisiae	Yes	Ethanol	SSF Batch
Oat spelt xylan	No	2.5%	Debaromyces hansenii	Yes	Ethanol	SSF
Wheat bran hemicellulose						Batch
Barley hull	Ammonia	3% w/v Glucan	Recombinant E. coli (KO11)	No	Ethanol	SSCF Batch
Xylan	Xylanase	13.1 g L^{-1} xylose	Clostridium butyricum CGS5	Yes	Hydrogen	SSF two-stage process batch-dark fermentation
Rice straw	NaOH	9.2 g L^{-1} xylose		Yes		
Wheat straw	Steam	9% WIS	Saccharomyces cerevisiae, TMB3400	Yes	Ethanol	SSF Fed-batch
Corn stover	Lactic acid and/or acetic acid	2% w/v Glucan	De Danske Spritfabrikker A/S	No	Ethanol	SSF batch
Wheat straw	Lime	6.3% Dry matter	B. coagulans DSM 2314	Yes	Lactic acid	SSF Fed-batch
Sweet sorghum bagasse	Ammonia fiber expansion	1% Glucan	Saccharomyces cerevisiae 424A	No	Ethanol	SSF
Switchgrass	Ammonia fiber expansion	4% of Cellulases	Saccharomyces cerevisiae 424A	Yes	Ethanol	Two-step SSCF Fed-batch
Convert corn	Ammonia	3% w/v Glucan	Recombinant E. coli KO11	Yes	Ethanol	TPSSF
Corn stover	Diluted alkaline	10.5% WIS	Actinobacillus succinogenes	No	Succinic acid	SSF Batch
Xylan	No		Clostridium amygdalinum strain C9	Yes	Hydrogen	Dark fermentation batch
Switchgrass	Hydrothermolysis	41 g L^{-1} glucan	Kluyveromyces marxianus IMB4	Yes	Ethanol	SSF batch
Wheat straw	Lime		E. coli	Yes	Ethanol	SSF batch

being directed towards developing a stable, ethanol-tolerant, robust recombinant ethanologenic organism capable of tolerating common fermentation inhibitors generated during pre-treatment.[63]

7.5.4   Applications

7.5.4.1   Biomedical Applications

For biomedical purposes, considerable research has been done in the area of hemicelluloses due to the biocompatibility, biodegradability, and high stability of these compounds. Many hemicelluloses are already widely used in the pharmaceutical industry, while others are as yet relatively unexploited.[64] More specialized drugs and methods of drug delivery will be necessary to fulfill requirements, as progress in the pharmaceutical industry leads to increased demands on materials for specific applications. Lots of hemicelluloses are already commonly used in this field, and other less investigated hemicelluloses may find application in the future.[65]

7.5.4.2   Hemicellulose Films

A number of research papers have focused on more fundamental research topics, while many studies on hemicellulose films and coatings have described well-defined commercial targets. This is especially characteristic of some of the work on chemically modified hemicelluloses. In chemically modified hemicelluloses, functionalization itself can appear to be the objective.[67] From the resulting products, films have been cast in some cases, and the properties of these have been assessed. Table 7.3 summarizes past research in producing barrier films from hemicelluloses.

7.5.5   Summary

Pre-extraction and isolation of hemicelluloses and lignin followed by the production of value-added products, such as ethanol, sugar-based polyesters, other chemicals, and biopolymers offer a potential opportunity in an integrated lignocellulose biorefinery. Using efficient methods to isolate and purify hemicelluloses would be beneficial to increasing their utilization. For practical purposes, by enzymatic and/or fermentation routes, conversion of hemicellulose into value-added useful products holds strong promise for the use of a variety of non-utilized and under-utilized agricultural residues. Now the conversion of hemicellulosic substrates to fermentable sugars is problematic. In order to produce fermentable sugars, some of the emerging pre-treatment methods, such as alkaline peroxide and ammonia fiber explosion (AFEX), generate solubilized and partially degraded hemicellulosic biomass that needs to be treated further with enzymes or other means. With the proper mix of hemicellulases (an enzyme cocktail) tailored for each biomass

Table 7.3 Water vapor permeability (WVP), water vapor transfer rate (WVTR), and strength properties of edible hemicellulose films.[66]

Hemicellulose	Source	Modification	Test temperature (°C)	WVP (10^{-11} gm^{-1} Pa^{-1} s^{-1})	WVTR (10^{-3} gm^{-2} s^{-1})	Tensile strength (MPa)	Elastic modulus (MPa)
β-glucan	barley, oat	none	30	68–169	7.5–10.0	0.91–13.02	
xylan	birch, grass corncob	mixing with wheat gluten (major component)	23		6.9–9.3	1–9	10–200
arabinoxylan	corn hull	none	22	2.3–4.3	4.7	9.7–60.7 53.8	365–1320 1316
arabinoxylan	maize bran	emulsified films are generated with globules of stearic or palmitic acid, triolein, palm oil	25	11.8–15.2	2.71–3.52	6.4–8.8	25.84–59.2
arabinoxylan	maize bran	none		17.7	3.92	26.5	72.4
		grafting with fatty acids from fish oils on the formed film by plasma and electron beam treatment	25	10.9–19.1	2.86–3.89		
arabinoxylan	maize bran	none		20.5	4.45		
		grafting with stearyl (meth) acrylate on the formed film by plasma and/or electron beam treatment	25	6.8–17.2	1.50–3.77		
arabinoxylan	maize bran	none		17.7–20.5	3.92–4.45		
		mixing with sucroesters as emulsifiers and hydrogenated palm oil as hydrophobic phase	25	9.31–13.82			
xylan	cotton stalk	residual lignin 1% (w/w lignin/xylan)	20		3.14	0.76	0.08
					2.52–3.08	1.08–1.39	0.11–0.49

conversion, and minimizing the formation of inhibitory compounds for fermentative organisms with the development of a suitable pre-treatment method, this vast renewable resource can be utilized for production of fuels and chemicals by fermentation. Much research needs to be done in the future to develop efficient and cost-effective pre-treatment methods, enzymes for hemicellulose conversion at an industrial scale, robust efficient microorganisms to ferment hemicellulosic sugars simultaneously in a cost-competitive way, and methods for cost-effective recovery of fermentation products.

7.6   Conclusions and Suggestions

Recently, conversion of biomass to chemicals and materials has been a research hotspot. New and great progress on efficient separation and high value added application of the three main biomass components (cellulose, hemicellulose, and lignin) have been made.

1. Many methods have been developed to separate cellulose, hemicellulose, and lignin from biomass, including acid/alkaline hydrolysis, ammonia fiber expansion, organic solvent extraction, hot water and ionic liquid treatment, *etc*. Furthermore, ionic liquid separation technology has attracted a great deal of attention due to its high efficiency and low reaction temperature properties. However, an economical, environmentally friendly, and effective separation method is still a great challenge for future research.

2. Considerable effort has already been devoted to developing reactions and applications of lignin, especially on the decomposition of lignin to produce low molecular mass phenolic compounds and on the preparation of lignin-derived polymeric material. Based on the recent progress, improvement of reaction selectivity of special phenolic compounds and their efficient separation from lignin degradation products should be a research hotspot to be solved.

3. Cellulose has been considered as an attractive alternative to produce fuel, chemicals, and functional materials. In long-term research, the integrated production of fuel and fine commodity chemicals from cellulose would be a preferred choice. Several functional materials based on cellulose derivatives or cellulose graft copolymers have already been designed successfully; however, more work still needs to be done to make them more practical.

4. Enzymatic and/or fermentation routes are selected as promising methods to convert hemicellulose into value-added useful products. However, much research needs to be done, including developing efficient enzymes for hemicellulose conversion at an industrial scale, cultivating robust microorganisms to ferment hemicellulosic sugars simultaneously in a cost-competitive way, and creating methods for cost-effective recovery of fermentation products.

Acknowledgements

The author thanks PhD students Shimin Kang, Yan Li, Shuai Peng, Aili Ma, Jun Ye, and Xianwei Zheng for their work in preparing this book chapter, and acknowledges financial support from the National Basic Research Program of China (973 Program) (No. 2013CB228104 and 2010CB732205), Doctoral Fund of Ministry of Education (20120172110011).

References

1. D. Mohan, C. U. Pittman, Jr. and P. H. Steele, *Energy Fuel*, 2006, **20**, 848.
2. J. Pérez, J. Munoz-Dorado, T. De la Rubia and J. Martinez, *Int. Microbiol.*, 2002, **5**, 53.
3. J. Bidlack, M. Malone and R. Benson, *Proc. Oklahoma Acad. Sci.*, 1992, 72, 51.
4. K. V. Sarkanen and C. H. Ludwig, *Lignins: Occurrence, Formation, Structure and Reactions*, Wiley-Interscience, New York, 1971.
5. J. McCarthy and A. Islam, *Lignin: Historical, Biological and Materials Perspectives*, ed. W. G. Glasser, R. A. Northey and T. P. Schultz, *ACS Symposium Series 742*, American Chemical Society, Washington DC, 2000, p. 2.
6. B. Saake and R. Lehnen, *Lignin: Ullmann's Encyclopedia of Industrial Chemistry*, Wiley-VCH, Weinheim, 2007, DOI: 10.1002/14356007.a15_305.pub3.
7. M. Kleinert and T. Barth, *Chem. Eng. Technol.*, 2008, **31**, 736.
8. S. Kang, X. Li, J. Fan and J. Chang, *Ind. Eng. Chem. Res.*, 2011, **50**, 11288.
9. P. Mäki-Arvela, T. Salmi, B. Holmbom, S. Willför and D. Y. Murzin, *Chem. Rev.*, 2011, **111**, 5638.
10. X. Li, E. Mupondwa, S. Panigrahi, L. Tabil, S. Sokhansanj and M. Stumborg, *Renew. Sust. Energ. Rev.*, 2012, **16**, 2954.
11. B. Kamm, P. R. Gruber and M. Kamm, *Biorefineries – Industrial Processes and Product: Status Quo and Future Directions*, Wiley-VCH, Germany, 2006, pp. 1, 16, 296.
12. C. L. Li, B. Knierima, C. Manisseri, *et al.*, *Bioresour. Technol.*, 2010, **101**, 4900.
13. P. Kumar, D. M. Barrett, M. J. Delwiche and P. Stroeve, *Ind. Eng. Chem. Res.*, 2009, **48**, 3713.
14. F. Teymouri, L. L. Perez, H. Alizadeh and B. E. Dale, *Appl. Biochem. Biotechnol.*, 2004, **113–116**, 951.
15. C. Liu and C. E. Wyman, *Appl. Biochem. Biotechnol.*, 2004, **113–116**, 977.
16. X. J. Pan, D. Xie, R. W. Yu and J. N. Saddler, *Biotechnol. Bioeng.*, 2008, **101**, 39.
17. H. Tadesse and R. Luque, *Energy Environ. Sci.*, 2011, **4**, 3913.
18. H. Yu, J. Hu and J. Chang, *Ind. Eng. Chem. Res.*, 2011, **50**, 7513.
19. C. Z. Li, Q. Wang and Z. K. Zhao, *Green Chem.*, 2008, **10**, 177.
20. T. Marzialetti, M. B. V. Olarte, *et al.*, *Ind. Eng. Chem. Res.*, 2008, **47**, 7131.

21. T. Stahlberg, M. G. Sorensen and A. Riisager, *Green Chem.*, 2010, **12**, 321.
22. S. S. Y. Tan, D. R. MacFarlane, *et al.*, *Green Chem.*, 2009, **11**, 437.
23. D. C. Dibble, C. L. Li, L. Sun, *et al.*, *Green Chem.*, 2011, **13**, 3255.
24. C. H. Zhou, X. Xia, D. S. Tong, *et al.*, *Chem. Soc. Rev.*, 2011, **40**, 5588.
25. D. M. Alonso, S. G. Wettstein, *et al.*, *Energy Environ. Sci.*, 2013, **6**, 76.
26. X. W. Zheng, Y. Fu, J. Chang, *et al.*, *Bioenerg. Res.*, 2013, DOI: 10.1007/ s12155-013-9298-3.
27. M. P. Garcia, A. Chaala, *et al.*, *Biomass Bioenerg.*, 2007, **31**, 222.
28. J. Zakzeski, P. C. A. Bruijnincx, A. L. Jongerius and B. M. Weckhuysen, *Chem. Rev.*, 2010, **110**, 3552.
29. P. Azadi, O. R. Inderwildi, R. Farnood and D. A. King, *Renew. Sust. Energ. Rev.*, 2013, **21**, 506.
30. M. Foston, G. A. Nunnery, X. Meng, Q. Sun, F. S. Baker and A. Ragauskas, *Carbon*, 2013, **52**, 65.
31. Q. Shen, T. Zhang, W. Zhang, S. Chen and M. Mezgebe, *J. Appl. Polym. Sci.*, 2011, **121**, 989.
32. M. Funaoka, *React. Funct. Polym.*, 2013, **73**, 396.
33. T. Saito, R. H. Brown, M. A. Hunt, D. L. Pickel, J. M. Pickel, J. M. Messman, F. S. Baker, M. Keller and A. K. Naskar, *Green Chem.*, 2012, **14**, 3295.
34. T. Qu, W. Guo, L. Shen, J. Xiao and K. Zhao, *Ind. Eng. Chem. Res.*, 2011, **50**, 10424.
35. W. Doherty, P. Mousavioun and C. Fellows, *Ind. Crop. Prod.*, 2011, **33**, 259.
36. A. G. Barneto, J. A. Carmona, A. Galvez and J. A. Conesa, *Energ. Fuels*, 2009, **23**, 951.
37. L. P. Christopher, *Integrated Forest Biorefineries: Challenges and Opportunities*, RSC Green Chemistry, www.rsc.org, 2013, 18, Chapter 1, p. 37.
38. G. Wang, W. Li, B. Q. Li and H. K. Chen, *Fuel*, 2008, **87**, 552.
39. R. Beauchet, F. Monteil-Rivera and J. M. Lavoie, *Bioresour. Technol.*, 2012, **121**, 328.
40. Y. Ye, Y. Zhang, J. Fan and J. Chang, *Bioresour. Technol.*, 2012, **118**, 648.
41. D. Klemm, B. Heublein, H. P. Fink and A. Bohn, *Angew. Chem. Int. Ed.*, 2005, **44**(22), 3358.
42. J. Kim and S. Yun, *Macromolecules*, 2006, **39**(12), 4202.
43. H. L. Kang, R. G. Liu and Y. Huang, *Polym. Int.*, 2013, **62**(3), 338.
44. D. Shen, R. Xiao, S. Gu and K. Luo, *RSC Adv.*, 2011, **1**, 164.
45. C. H. Zhou, X. Xia, C. X. Lin and D. S. Tong, *Chem. Soc. Rev.*, 2011, **40**, 5588.
46. D. M. Alonso, S. G. Wettstein and J. A. Dumesic, *Chem. Soc. Rev.*, 2012, **41**, 8075.
47. Y. Sun and J. Y. Cheng, *Bioresour. Technol.*, 2002, **83**(1), 1.
48. C. H. Y. Chang and L. N. Zhang, *Carbohydr. Polym.*, 2011, **84**(1), 40.
49. M. W. Frey, *Polym. Rev.*, 2008, **48**(2), 378.
50. C. H. W. Kim, D. S. Kim, S. Y. Kang, M. Marquez and Y. L. Joo, *Polymer*, 2006, **47**, 5097.

51. H. S. Qi, X. F. Sui, J. Y. Yuan, Y. Wei and L. N. Zhang, *Macromolecules*, 2010, **295**, 695.
52. M. Gericke, J. Trygg and P. Fardim, *Chem. Rev.*, DOI: 10.1021/cr300242j.
53. X. G. Luo and L. N. Zhang, *Food Res. Int.*, 2013, **52**, 387.
54. D. Roy, M. Semsarilar, J. T. Guthrie and S. Perrier, *Chem. Soc. Rev.*, 2009, **38**, 2046.
55. M. S. Izydorczyk and J. E. Dexter, *Food Res. Int.*, 2008, **41**, 850.
56. A. X. Jin, J. L. Ren, F. Peng, F. Xu, G. Y. Zhou and R. C. Sun, *Carbohydr. Polym.*, 2009, **78**, 609.
57. M. J. Jin, M. W. Lau, V. B. Balan and B. E. Dale, *Bioresour. Technol.*, 2010, **101**, 8171.
58. F. K. Kazi, J. A. Fortman, R. P. Anex, D. D. Hsu, A. Aden, A. Dutta, *et al.*, *Fuel*, 2010, **89**, S20.
59. T. Q. Yuan, F. Xu, J. He and R. C. Sun, *Biotechnol. Adv.*, 2010, **28**, 583.
60. F. Xu, Z. C. Geng, J. X. Sun, C. F. Liu, J. L. Ren, R. C. Sun, *et al.*, *Carbohydr. Res.*, 2006, **341**, 2073.
61. F. Xu, C. F Liu, Z. C. Geng, J. X. Sun, R. C. Sun, B. H. Hei, *et al.*, *Polym. Degrad. Stab.*, 2006, **91**, 1880.
62. F. Xu, J. X. Sun, Z. C. Geng, C. F. Liu, R. C. Sun, P. Fowler, *et al.*, *Carbohydr. Polym.*, 2007, **67**, 56.
63. F. Xu, R. C. Sun, M. Z. Zhai, J. X. Sun, D. She, Z. C. Geng, *et al.*, *Sep. Sci. Technol.*, 2008, **43**, 3351.
64. J. Xu, M. H. Thomsen and A. B. Thomsen, *J. Biotechnol.*, 2009, **139**, 300.
65. Q. Xu, Y. L. Chao and Q. B. Wang, *Carbohydr. Polym.*, 2009, 77, 435.
66. F. Peng, P. Peng, F. Xu and R. C. Sun, *Biotechnol. Adv.*, 2012, **30**, 879.
67. M. L. H. Natanya and P. David, *Biomacromolecules*, 2008, **9**, 1493.

CHAPTER 8

Bioactive Compounds from Biomass

CHUN-PING XU*[a] AND RICK ARNEIL D. ARANCON[b]

[a] College of Food and Biological Engineering, Zhengzhou University of Light Industry, Zhengzhou, Henan, 450002, P.R. China; [b] Departamento de Quimica Organica, Universidad de Cordoba, Campus de Rabanales, Edificio Marie Curie (C-3), Ctra Nnal IV, Km 396, E-14014 Córdoba, Spain
*Email: c.p.xu@zzuli.edu.cn

8.1   Value of Biomass

8.1.1   Introduction

Biomass, a renewable energy source, is a biological material from living, or recently living, organisms. The solar energy is usually stored in plant organisms as carbohydrates through photosynthesis. Since other animals usually eat plants and consume the products of photosynthesis, it can be safely put that all living things on earth rely on this stored energy from the sun. Biomass includes a good number of common products, such as crops (especially energy crops, including corn, cassava, wheat, sugar cane, sugar beet, soybeans, rapeseed, sunflower, *etc.*), herbaceous or woody plants (including high-fiber plants, such as miscanthus, switchgrass, hybrid poplar), agriculture, forestry, animal husbandry waste (such as corn stalks and leaves ear, bagasse, sawdust, rice husk, coconut shell, *etc.*) or industrial organic waste, waste oil, fruit and vegetable waste, even more extensively, including biogas or methane hydrates (methane hydrate) and other resources.[1]

RSC Green Chemistry No. 27
Renewable Resources for Biorefineries
Edited by Carol Sze Ki Lin and Rafael Luque
© The Royal Society of Chemistry 2014
Published by the Royal Society of Chemistry, www.rsc.org

At present, mankind has to face the greenhouse effect, environmental pollution, and the lack of fossil fuels, which are primary sources of both energy and chemical raw materials. Thus, biomass utilization and conversion into industrial chemicals has become an important research field. Research on biomass does not only focus on the maximization of the conversion of biomass into useful chemicals, but an equal attention is also given to the development of green conversion technologies.[2,3]

Biomass is an important renewable resource and is essentially considered in-exhaustible, like wind, solar, and geothermal energy. Moreover, growing crops, the most common biomass source, can also absorb carbon dioxide and, if utilized heavily, can reduce greenhouse gas accumulation. The biomass biotechnology can be divided into three major directions, namely: (1) bioenergy, including various forms of energy such as bio-diesel, bio-ethanol, bio-hydrogen production systems, and energy produced from burning biomass; (2) biomaterials, such as the transformation of biomass into functional polymer materials (polyoxyalkanoates, PHA); and (3) biomass refinery which is mainly about the conversion of biomass to produce chemicals for food, cosmetics, and pharmaceutical applications.[3–5]

This chapter reviews the main chemical groups of bioactive compounds in plant biomass, taking tobacco biomass as a case study, for use in the food, cosmetics, pharmaceutical, and other related industries.

8.1.2 Screening of Bioactive Activity from Biomass

Natural products have been one of the major resources for chemical diversity in the pharmaceutical industry.[6] Although many drugs are made by methods of chemical synthesis, most of the scaffolds or core structures for chemical synthesis are based on isolated natural products. The key advantage of natural products over synthetic compounds is that natural products have more chemical diversity with steric structures and greater potent bioactivities due to natural selection than synthetic compounds. Hence, taking into account the great biodiversity of plant species, the appropriate screening methodology from different plant biomass sources for bioactive compounds is of great interest. To design the screening methodology, many criteria need to be considered. These criteria include, among others, the nature of the sought-after bioactive compounds (in terms of molecular weight, molecular structure, thermal stability, or solubility) and the corresponding bioactivity sought.[7] Figure 8.1 shows a proposed workflow for the screening of bioactive compounds from biomass. The efficient extraction method should be selected according to the predicted nature of the target bioactive compounds.[8,9] The bioactivities of different compounds obtained by diverse extraction conditions should be examined by performing the various functional activity assays, for instance, antimicrobial activity assays, antioxidant activity assays, or antihypertensive activity assays.[10] Once the target biological activities have been confirmed, the structure and chemical characterization of the bioactive components needs to be elucidated. It is

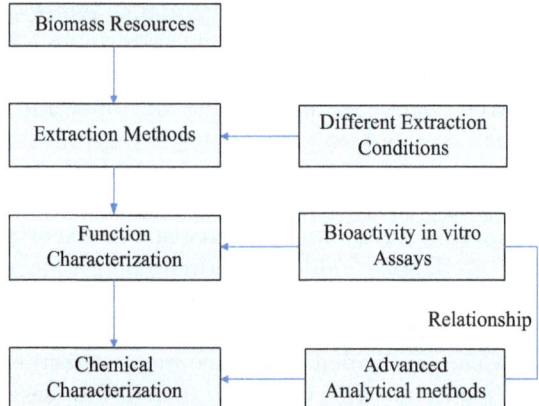

Figure 8.1 Basic scheme showing the proposed workflow for the screening of bioactive compounds from biomass.

always necessary to demonstrate the relationship between the biological activities and defined chemical structure.[11,12]

8.2   Bioactive Compounds from Biomass

Bioactive compounds in plant biomass are classified according to different criteria. One classification is based on biological activity and clinical function of compounds. This classification is complicated because of the fact that the clinical outcome is not exclusively related to chemically related compounds; even chemically very different molecules might produce similar bioactivities. Another classification is based on families and genera of the plants producing similar bioactive compounds, as closely related plant species often produce chemically similar bioactive compounds. However, there are also some examples that even genetically different plant species produce similar bioactive compounds. Since the main focus of this chapter is the bioactive chemical compounds of plants, it is useful to categorize them according to the combination of biochemical pathways and chemical classes. The following section is a brief presentation of the main chemical groups of bioactive compounds in plants. Some compounds are highlighted and classified as a group because of their importance in application.

8.2.1   Carbohydrates

Carbohydrates account for the greatest proportion of plant biomass, with cellulose and hemicellulose acting as major structural components of plant tissues and cell walls. Pectic polysaccharides from plants have been the subject of studies for long time mainly focused on their physical properties, their chemical and physical modification, and their application. These polysaccharides have the potential for use as prebiotics as they are not easily digested in the human gut and can be considered as a rich source of dietary fiber. These polysaccharides have also been shown to have immunomodulating, anticancer, anti-inflammatory, antiviral, and antioxidant activities. The bioactivities might differ depending on the degree of sulfonation, molecular weight, predominant monosaccharide content, and glycosidic branching.[13,14]

8.2.2 Peptides and Proteins

Proteins from plant organisms have received attention recently due to their potential bioactive and functional properties. Plant proteins (mainly from vegetables) also have been receiving more attention due to their potential applications in the cosmetics industry.[15] In addition, there is also interest in the development of special proteins in plants for healthcare, such as antibodies, antigens, and vaccines. Considering their application and economic importance, development of the methods to improve protein extraction is also the subject of current research and development.[16]

8.2.3 Secondary Metabolites

Bioactive compounds in plants often have pharmacological or toxicological effects in humans and other animals. Because they are often produced as secondary metabolites, bioactive compounds in plants can be defined as secondary plant metabolites having pharmacological or toxicological effects.[17]

8.2.3.1 Polyphenols

Polyphenols are considered secondary metabolites of plants involved in the chemical defense of plants against predators and in plant–plant interactions. Polyphenols are found in virtually all families of plants, and comprise up to 50% of the dry weight of leaves. The main activity related to phenolic compounds is antioxidant activity. In addition to their strong antioxidant activities, plant polyphenols are known to possess other activities including antibacterial, chemopreventive, UV-protective, anticancer, and anti-inflammatory effects, act as detoxifying agents against heavy metals, and have myriad other bioactivities that could potentially be exploited for application in functional foods.[18–22]

8.2.3.2 Carotenoids

Carotenoids, a family of natural pigments widespread in nature, have attracted many researchers all over the world because of their commercially desirable properties, such as their natural origin, high versatility, and null toxicity. They also provide both lipophilic and hydrophilic colorants such as provitamin A. Carotenoids occur in many fruits and vegetables, including tomatoes, carrots, watermelons, kale, and spinach. They are also powerful antioxidants, protecting the cells of the body from damages caused by free radicals.[23] Carotenoids are thought to be active agents for the prevention of cancer,[24] cardiovascular diseases,[25] and muscular degeneration.[26]

8.2.3.3   Glycosides

Glycosides consist of various categories of secondary metabolites bound to an oligosaccharide or to uronic acid in plants. Almost all plants store chemicals in the form of inactive glycosides which are activated by hydrolytic cleavage into a sugar and a non-sugar component (aglycone). The naming of glycosides is specifically based on the type of sugar attached: glucoside (glucose), pentoside (pentose), fructoside (fructose), *etc.* Many such plant glycosides are used as medications. Cardiac glycosides occur in certain plants, for example *Digitalis, Strophanthus, Urginea,* and act on the contractile force of cardiac muscles; some are used as cardiotonics and anti-arrhythmics.[27] Digitalis glycosides are a group of cardiotonic and antiarrhythmic glycosides derived from *Digitalis purpurea* and *D. lanata,* or any drug chemically and pharmacologically related to these glycosides.[28]

8.2.3.4   Tannins

Tannin is a general name for complex phenolic substances that are capable of tanning animal hides into leather by binding collagen. Tannins are one of the most abundant secondary metabolites made by plants, commonly ranging from 5% to 10% dry weight of plant leaves.[29,30] Tannins are available in two forms from plants: condensed and hydrolysable. Condensed tannins are large polymers of flavonoids while hydrolysable tannins are polymers composed of a monosaccharide core (usually glucose) with several catechin derivatives attached.[31,32] Tannins can defend leaves against insect herbivores by deterrence and/or toxicity. Examples of plant families associated with the presence of tannins are Fagaceae (beech family) and Polygonaceae (knotweed family).[33,34] In recent years, many researchers demonstrated that tannins have positive effects on animals by antimicrobial, anti-helmintic protein by-passed effects in ruminants and astringents in cases of diarrhea, skin bleeding, and transudates.[35]

8.2.3.5   Terpenoids

Sometimes called isoprenoids, terpenoids are a large and diverse class of naturally occurring organic chemicals similar to terpenes, derived from five-carbon isoprene units. Plant terpenoids are used extensively for their aromatic qualities. Just like terpenes, the terpenoids can be classified according to the number of isoprene units. They play an important role in traditional herbal remedies and are under investigation for antibacterial, anti-neoplastic, and other pharmaceutical functions.[36]

8.2.3.6   Phenylpropanoids

Phenylpropanoids represent a large group of natural products containing a phenyl ring attached to a three-carbon propane side chain (C_6–C_3) in their structure.[37] Phenylpropanoids are found throughout the plant kingdom and provide protection from ultraviolet light, mediate plant–pollinator interactions as floral pigments and scent compounds, and defend against herbivores and pathogens.[38] Their functions vary greatly, a range of which have been utilized in herbal remedies as antioxidants, anticancer, antiviral, anti-inflammatory, and antibacterial agents.[39]

8.2.3.7 Resins

'Resin' is used to refer to hydrocarbon secretions of many plants, particularly coniferous trees. The resins are complex lipid-soluble mixtures of both non-volatile and volatile compounds. The non-volatile fraction consists of diterpenoid and triterpenoid compounds, while the volatile fraction is mostly composed of mono- and sesquiterpenoids. Resins are valued for their chemical properties and associated uses, such as the production of varnishes, adhesives, and food-glazing agents.[40]

8.2.3.8 Lignans

Lignans are composed of two phenylpropanoid units joined together to form an 18-carbon skeleton, with various functional groups connected. They are always found in woody tissues and resins of plants. Several lignans show clinical activity, such as phytoestrogenic, cathartic, antineoplastic, antiviral, and liver protective activities.[41]

8.2.3.9 Alkaloids

Alkaloids are a group of chemical compounds containing basic nitrogen atoms with a great structural diversity, and there is no uniform classification of alkaloids. They are of limited distribution in the plant kingdom. The various groups of alkaloids have diverse clinical properties, such as antimicrobial, anti-hyperglycemic, anti-inflammatory, and pharmacological effects.[42]

8.2.3.10 Summary

As shown in this section, biomass-related compounds can be a good functional source that may be used as ingredients in various industries.

8.3 Case Study: Bioactive Compound from Tobacco Waste

Tobacco (*Nicotiana* sp.) is the most widely grown crop in the world and is cultivated in more than 130 countries. Tobacco biomass waste is classified as an agro-industrial waste. In 2005, the total production of global tobacco waste was more than 1.25 million metric tons. In China alone, 460 million kilograms of tobacco waste per year is generated at various stages of tobacco post-harvest processing and during the manufacture of tobacco-derived products.[43] Due to the presence of residual nicotine, tobacco waste is toxic and thus legislation has been enforced for the controlled disposal of tobacco waste to avoid its harmful effects. At present, the majority of the waste is destroyed by burning or burying. As a type of high organic biomass, tobacco wastes have been used as desulfurization adsorbents and a tailored organic fertilizer.[44,45]

The chemical composition of tobacco leaves and biomass waste has received a considerable amount of attention in recent years, and tobacco waste is considered to be a good source of a large number of bioactive substances.[46,47] Tobacco waste contains a variety of valuable chemical constituents, such as nicotine, solanesol, sclareol, vitamin E, riboflavin, tobacco protein, *etc*. They can be obtained by a series of chemical extraction and refining methods. The bioactive compounds can be used in pharmaceutical, chemical, and other industries.[48] In the following sections, the active compounds from tobacco waste are reviewed by metabolite type.

8.3.1   Sclareol

Sclareol (Figure 8.2), abundant in *Salvia sclare L.* and *Nicotiana glutinosa*, is a diterpene natural product of high value for the fragrance industry. Its labdane carbon skeleton and its two hydroxyl groups also make it a valued building block for other chemicals. Sclareol has many bioactivities, such as anti-inflammatory, anti-infection, and anti-cancer activity with lower toxicity.[49]

8.3.2   Alkaloids

Over 40 kinds of alkaloids from tobacco have been isolated and identified. The major alkaloids from tobacco are nicotine, cotinine, nornicotine, myosmine, nicotyrine, anabasine, and anatabine. The nicotine content is highest of most of the alkaloids and nicotine (Figure 8.3) has insecticidal, detoxifying, and refreshing activities. Nicotine is also commonly used to synthesize niacin and niacinamide (key intermediates). Niacinamide (Figure 8.3) is the amide of nicotinic acid (vitamin B3/niacin), a water-soluble

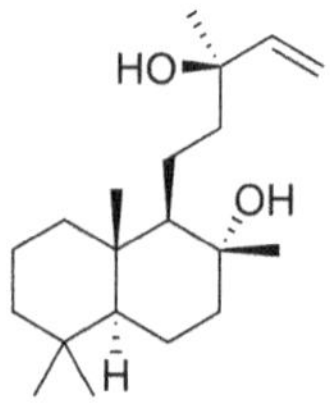

Figure 8.2   Structure of sclareol.

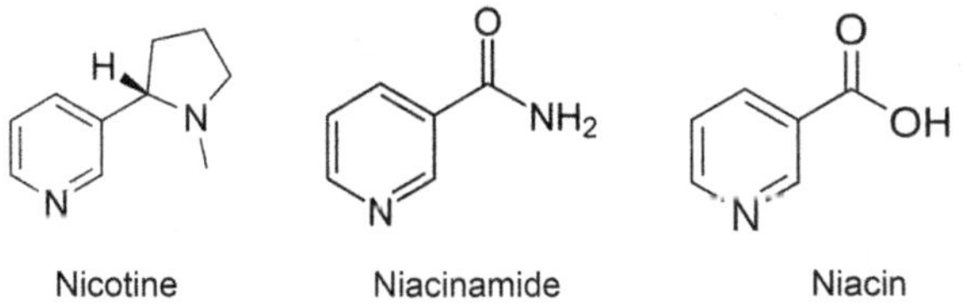

Figure 8.3   Structures of nicotine, niacinamide and niacin.

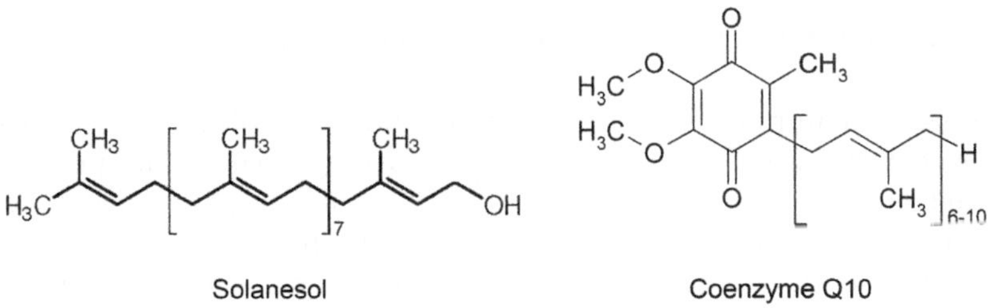

Figure 8.4   Structures of solanesol and coenzyme Q10.

vitamin that is part of the vitamin B group. Niacin (Figure 8.3), on the other hand, is one of the vitamins associated with a pandemic deficiency disease: niacin deficiency (pellagra). Niacin has been used for over half a century to increase levels of HDL in the blood and has been found to modestly decrease the risk of cardiovascular events.[50]

8.3.3   Solanesol

Solanesol (Figure 8.4) is a polyisoprenoid alcohol mainly found in the surface of tobacco leaves. Solanesol is mainly present in flue-cured tobacco with the content of 0.5% to 1% (w/w).[51] In addition to being sold as a cosmetic ingredient in its unmodified form, solanesol is also an intermediate compound in the synthesis of many high-value biochemicals such as vitamin-K analogs and coenzyme-Q10 (ubiquinone). Coenzyme Q10 (Figure 8.4) is used successfully in treating ischemic heart disease, chronic heart failure, toxin-induced cardiomyopathy, hypertension, and hyperlipidemia. Among the increasing number of pharmacological uses ascribed to coenzyme Q10 are anticancer (in particular breast cancer) activity, treatment of periodontal disease, diabetes, Parkinson's, Alzheimer's, Huntington's disease, and to counteract the aging process.[52]

8.3.4   Proteins and Amino Acids

The proteins from tobacco are highly unusual in the plant kingdom in that they are complete and well balanced in their amino acid content. This amino acid content reaches about 20.5% of the dry weight of tobacco. There are over 43 kinds of amino acids found in tobacco, including high levels of eight

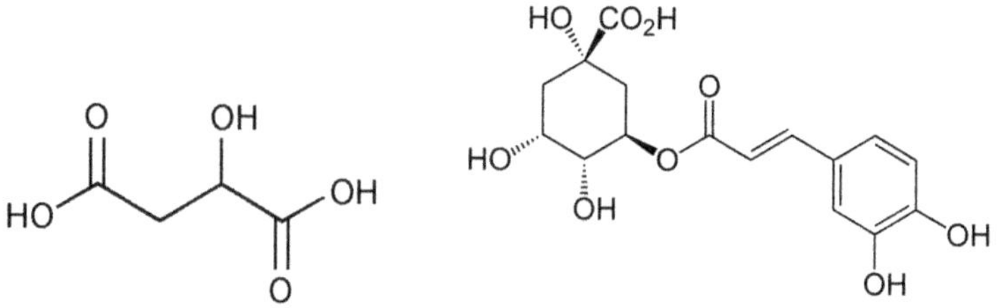

Figure 8.5   Structures of malic acid and chlorogonic acid.

of the essential amino acids for the human body (glutamic acid, acetic acid, proline acid, arginine, histidine, alanine, aspartic acid, *etc.*). Plant protein is an excellent natural nutritional food ingredient and food additive.[53] Even some amino acids have a therapeutic role, for example aspartate has a potential adverse side effect for cardiac insufficiency.[54] The tobacco plant protein can not only become a substitute for animal protein, especially for lactose-intolerant patients, but can also be added to the products of meat, egg, fish, and cakes to increase their nutritional value.[55]

8.3.5   Organic Acids

Tobacco is also rich in organic acids, mainly malic acid, citric acid, oxalic acid and malonic acid, *etc.* A substantial portion of such acids are complexed as salts with nicotine, ammonia, and inorganic anions of calcium, potassium, and sodium. The content of malic acid (Figure 8.5) is higher than other organic acids, and can be used to synthesize poly(α-hydroxyalkanoate) for bone repair and muscle regeneration.[56] Chlorogonic acid (Figure 8.5) is a member of the caffeoylquinic acids family, which possess a wide range of biological activities, such as antibacterial, antioxidant, antimutagenic, hepatocyte protective, and inhibitory of HIV-1 RT, the human herpes simplex virus, adenoviruses, SARS, and AIV (H5N1).[48]

8.3.6   Carbohydrates

Tobacco contains a lot of polysaccharides, including cellulose, hemicellulose, lignin, *etc.* The carbohydrate content can reach to about 22% of the dry weight of tobacco. The polysaccharides can be used as a raw material for chemical building blocks, such as fumaric acid, succinic acid, *etc.* A recent report indicated that tobacco polysaccharides have antioxidant activity.[57]

References

1. G. W. Huber, S. Iborra and S. Corma, *Chem. Rev.*, 2006, **106**, 4044.
2. P. McKendry, *Bioresour. Technol.*, 2002, **83**, 37.
3. P. McKendry, *Bioresour. Technol.*, 2002, **83**, 47.

4. D. Mohan, C. U. Pittman and P. H. Steele, *Energ. Fuel.*, 2006, **20**, 848.
5. M. E. Himmel, S. Y. Ding, D. K. Johnson, W. S. Adney, M. R. Nimlos, J. W. Brady and T. D. Foust, *Science*, 2007, **315**, 804.
6. M. M. Cowan, *Clin. Microbiol. Rev.*, 1999, **12**, 564.
7. T. O. Larsen, J. Smedsgaard, K. F. Nielsen, M. E. Hansen and J. C. Frisvad, *Nat. Prod. Rep.*, 2005, **22**, 672.
8. P. Claeson, U. Göransson, S. Johansson, T. Luijendijk and L. Bohlin, *J. Nat. Prod.*, 1998, **61**, 77.
9. Y. Sudjaroen, R. Haubner, G. Würtele, W. E. Hull, G. Erben, B. Spiegelhalder, S. Changbumrung, H. Bartsch and R. W. Owen, *Food Chem. Toxicol.*, 2005, **43**, 1673.
10. R. J. P. Cannell, *Natural Products Isolation*, Springer, 1998, 4.
11. R. Bhadra, R. J. Spanggord, D. G. Wayment, J. B. Hughes and J. V. Shanks, *Environ. Sci. Technol.*, 1999, **33**, 3354.
12. J. W. Jaroszewski, *Planta Med.*, 2005, **71**, 795–802.
13. S. Wasser, *Appl. Microbiol. Biotechnol.*, 2002, **60**, 258.
14. C. K. Wong, K. N. Leung, K. P. Fung and Y. M. Choy, *J. Int. Med. Res.*, 1994, **22**, 299.
15. M. Simon and F. Azam, *Mar. Ecol. Prog. Ser.*, 1989, **51**, 201.
16. M. W. Hunkapiller, E. Lujan, F. Ostrander and L. E. Hood, *Method. Enzymol.*, 1982, **91**, 227.
17. P. Gantet and J. Memelink, *Trends Pharmacol. Sci.*, 2002, **23**, 563.
18. C. A. Rice-Evans, N. J. Miller and G. Paganga, *Free Radical Bio. Med.*, 1996, **20**, 933.
19. L. Bravo, *Nutr. Rev.*, 1998, **56**, 317.
20. C. Rice-Evans, N. Miller and G. Paganga, *Trends Plant Sci.*, 1997, **2**, 152.
21. C. A. Rice-Evans, N. J. Miller, P. G. Bolwell, P. M. Bramley and J. B. Pridham, *Free Radical Res.*, 1995, **22**, 375.
22. K. E. Heim, A. R. Tagliaferro and D. J. Bobilya, *J. Nutr. Biochem.*, 2002, **13**, 572.
23. R. Edge, D. J. McGarvey and T. G. Truscott, *J. Photochem. Photobiol. B*, 1997, **41**, 189.
24. N. Hoyoku, M. Murakoshi, T. Ii, M. Takemura, M. Kuchide, M. Kanazawa, X. Y. Mou, S. Wada, M. Masuda, Y. Ohsaka, S. Yogosawa, Y. Satomi and K. Jinno, *Cancer Metast. Rev.*, 2002, **21**, 257.
25. G. S. Omenn, G. E. Goodman, M. D. Thornquist, J. Balmes, M. R. Cullen, A. Glass, J. P. Keogh, F. L. Meyskens, B. Valanis, J. H. Williams, S. Barnhart and S. Hammar, *New Engl. J. Med.*, 1996, **334**, 1150.
26. L. J. Popplewell, A. Abu-Dayya, T. Khanna, M. Flinterman, N. A. Khalique, L. Raju, C. L. Øpstad, H. R. Sliwka, V. Partali, G. Dickson and M. D. Pungente, *Molecules*, 2012, **17**, 1138.
27. K. S. Lee and W. Klaus, *Pharmacol. Rev.*, 1971, **23**, 193.
28. B. Noé, B. Hagenbuch, B. Stieger and P. J. Meier, *Proc. Natl. Acad. Sci. USA*, 1997, **94**, 10346.
29. R. V. Barbehenn and C. C. Peter, *Phytochemistry*, 2011, **72**, 1551.
30. A. Scalbert, *Phytochemistry*, 1991, **30**, 3875.

31. L. Wolf Bors, F Yeap, H. Norbert, M. Christa and S. Kurt, *Antioxidants Redox Signaling*, 2001, **3**, 995.
32. S. Hassanpour, N. Maherisis, B. Eshratkhah and F. B. Mehmandar, *Int. J. Forest, Soil and Erosion*, 2011, **1**, 47.
33. W. Y. Huang, Y. Z. Cai, J. Xing, H. Corke and M. Sun, *Planta Med.*, 2008, **74**, 43.
34. Y. Moumou, F. Trotin, J. Vasseur, G. Vermeersch, R. Guyon, J. Dubois and M. Pinkas, *Planta Med.*, 1992, **58**, 516.
35. L. Bravo, *Nutr. Rev.*, 1998, **56**, 317.
36. J. Bohlmann, G. Meyer-Gauen and R. Croteau, *Proc. Natl. Acad. Sci. USA*, 1998, **95**, 4126.
37. J. T. Knudsen, L. Tollsten and L. G. Bergström, *Phytochemistry*, 1993, **33**, 253.
38. J. T. Knudsen, R. Eriksson, J. Gershenzon and B. Ståhl, *Bot. Rev.*, 2006, **72**, 1.
39. R. L. Nicholson and R. Hammerschmidt, *Annu. Rev. Phytopathol.*, 1992, **30**, 369.
40. M. A. Phillips and R. B. Croteau, *Trends Plant Sci.*, 1999, **4**, 184.
41. W. D. MacRae and G. H. Towers, *Phytochemistry*, 1984, **23**, 1207.
42. A. A. Watson, G. W. J. Fleet, N. Asano, R. J. Molyneux and R. J. Nash, *Phytochemistry*, 2001, **56**, 265.
43. X. Zhu, F. Zheng and Z. Cao, *Acta Tabacaria Sinica*, 2006, **12**, 58.
44. M. Seredych and T. J. Bandosz, *Environ. Sci. Technol.*, 2007, **41**, 3715.
45. S. Chaturvedi, D. K. Upreti, D. K. Tandon, A. Sharma and K. Dixit, *J. Environ. Biol.*, 2008, **29**, 759.
46. M. E. Hegazy, T. Hirata, A. Abdet-lateff, M. H. El-Razek, A. E. Mohamed, N. M. Hassan, P. W. Pare and A. A. Mahmoud, *Z. Naturforsch. A*, 2008, **63**, 403.
47. C. Crofcheck, M. Loiselle, J. Weekley, I. Maiti, S. Pattanaik, P. M. Bummer and M. Jayt, *Biotechnol. Prog.*, 2003, **19**, 680.
48. J. Wang, D. Lu, H. Zhao, B. Jiang, J. Wang, X. Ling, H. Chai and P. Ouyang, *J. Serb. Chem. Soc.*, 2010, **75**, 875.
49. G. J. Huang, C. H. Pan and C. H. Wu, *J. Natl. Prod.*, 2012, **75**, 54.
50. N. Nuchtavorn and L. Suntornsuk, *J. Chromatogr. Sci.*, 2012, **50**, 151.
51. J. Chen, J. Liu, F. S. C. Lee and X. Wang, *J. Sep. Sci.*, 2008, **31**, 137.
52. Y. Tian, T. Yue, Y. Yuan, P. K. Soma, P. D. Williams, P. A. Machado, H. Fu, R. J. Kratochvil, C. I. Wei and Y. M. Lo, *Bioresour. Technol.*, 2010, **101**, 7877.
53. S. D. Kung, J. A. Saunder, T. C. Tso, D. A. Vaughan, M. Womack, R. C. Staples and G. R. Beecher, *J. Food Sci.*, 1980, **45**, 320.
54. O. I. Pisarenko, *Clin. Exp. Pharmacol. Physiol.*, 1996, **23**, 627.
55. M. Parameswaran, S. R. Parmar, K. S. Prajapati and M. K. Chakraborty, *Plant Foods Hum. Nutr.*, 1988, **38**, 269.
56. V. Jeanbat-Mimaud, C. Barbaud, J. P. Caruelle, D. Barritault, S. Cammas-Marion, V. Langlois and P. Guerin, *J. Biomater. Sci. Polym. Ed.*, 2000, **11**, 979.
57. C. Xu, C. Yang and D. Mao, *Pharmacognosy Mag.*, 2014, **10**, 66.

Subject Index

Illustrations and figures are in **bold**. Tables are in *italics*.